MODERN PETROLEUM

A BASIC PRIMER OF THE INDUSTRY

MODERN PETROLEUM

A BASIC PRIMER OF THE INDUSTRY

Second Edition

BY BILL D. BERGER AND KENNETH E. ANDERSON

ASSISTED BY G. L. FARRAR AND KATHRYNE E. PILE

PennWell Books
PennWell Publishing Company
Tulsa, Oklahoma

This book is dedicated to Col. Edwin L. Drake
and all the men and women since his time who
have made the petroleum industry what it is today.

Copyright © 1978, 1981 by
PennWell Publishing Company
1421 South Sheridan Road/P.O. Box 1260
Tulsa, Oklahoma 74101

Library of Congress Cataloging in Publication Data

Berger, Bill D.
    Modern petroleum.

    Bibliography
    Includes index.
    1. Petroleum engineering.   I. Anderson, Kenneth E.
II. Title.
TN870.B457 1981   622'.338   74-53168
ISBN 0-87814-172-3   AACR2

Printed in the United States of America

1 2 3 4 5   85 84 83 82 81

Contents

Preface

It is not the intent of this book to go into great technical detail. Rather it has been designed to give a broad overview of all aspects of the petroleum industry. Hopefully, it will prove beneficial not only to new students of petroleum and new employees, but also to those already affiliated with the oil and gas industry who may not have had the opportunity to become familiar with operations outside their own sphere of interest. It is also hoped this book may prove to be of value to all who may be interested in learning where and how we obtain our major source of energy today.

Such a project would not be possible, of course, without the assistance and cooperation of a great many individuals and organizations. The authors are deeply indebted to Dr. Don Adams of Oklahoma State University, who read portions of the manuscript and made many helpful suggestions, and to Mr. Fred Adlam, Noble Drilling Company; Mr. Van McQueen, Tulsa Litho; Mr. Bruce Weyland, Petroleum Extension Service, University of Texas at Austin; and Mr. Greg Snyder and Mr. Rocky Hails, Snyder-Hails Studio.

Also, Mr. John Feehery, Amoco Torch; Mr. Frank D. Hampton, Cities Service Today; Mr. Joseph V. Maranto, Mobil World; Mr. Maury Bates, Conoco 77; Mr. Bruce Kleinman, Shell Ecolibrium; and Mr. Gilbert Grosvenor and Mr. William N. Palmstrom, National Geographic Magazine.

Mr. Gene T. Kinney, the Oil and Gas Journal; Mr. John C. McCaslin, PennWell Publishing Company; Ms. Barbara Pritchard, Society of Petroleum Engineers of AIME; Mr. Gary D. Howell, American Association of Petroleum Geologists; Mrs. Mavis Yoachum, University of Oklahoma Press; the Dresser-Atlas Division, Dresser Industries, Inc., and the Oklahoma Petroleum Council.

Mr. Robert O. Frederick, Drilling Associated Publishers, Inc.; Mr. J. H. Green, the American Petroleum Institute; The Atlantic Richfield Company; Mrs. Billie Linduff, Drumright, Oklahoma; Mrs. Leanna McKenny, Tiny's Acid and Frac Service; Mr. J. B. Red, Ketal Oil Producing Company, and many others.

Very special thanks are due our editor, Mr. G. L. Farrar, for his enthusiastic encouragement throughout this project, and to Mrs. Vicki Evans who typed the manuscript.

Bill D. Berger
Ken Anderson

CHAPTER 1

How It All Began

Oil. Petroleum. Black gold. Formed under the surface of the earth millions of years ago, it has long been known to man. Yet only during the past hundred years or so have we realized oil's value and usefulness. In hardly more than a century, our modern society has become almost totally dependent upon the petroleum industry. Today we use petroleum products for transportation, heating, electricity, fertilizers, fabrics, medicines, plastic goods, pesticides, paints, beverages, and thousands of other items.

The word *petroleum* is derived from the Greek and Latin words *petra,* meaning rock, and *oleum,* meaning oil. Thus, early petroleum samples were appropriately named "rock oil." This rock oil has been used for nearly 5,000 years—since recorded time began. The great civilizations of the ancient Middle East (the Babylonians, Assyrians, Persians, Ionians, and Sumerians) used bitumen or asphalt to caulk buildings and ships. The Chinese dug for natural gas and used it to light the Emperor's palace. They transported the gas through the first known pipeline: hollow bamboo sections. The Greeks also used natural gas; the spots where their prophets' oracles or fires appeared were really lighted seeps that burned endlessly.

By the time the Christian era began, petroleum had been recorded in the Bible and in Greek and Sumerian writings. As early as 100 A.D., alchemists described a way of distilling heavy oil, found in surface pools, to produce naphtha. Arabs and Syrians were burning pitch and naphtha for illumination, experiencing their Light Ages at the same time Europeans were in the Dark Ages. Later, naphtha was used by medieval soldiers as Greek Fire.

During these hundreds of years before the New World was discovered, the source of most of the petroleum was generally seeps at ground level or an underground pool close to the surface that was found when digging or boring for brine or water. Although the earliest petroleum products were limited in uses, the New World opened up a vast new market for oil and gas.

THE NEW WORLD

When the first Spanish explorers ventured into the newly discovered continent, they found oily marshes and tar pools in Trinidad, Venezuela, Mexico, and later in California. Mexican natives called the oily substance *chapapote,* also known as *mene* in Venezuela and other parts of Latin America. Precolombian Indians used it as medicine, pitch, glue, illuminating fluid, ointment, and incense. As oceanic voyages to the Indies became more frequent, sailors learned to use heated heavy petroleum residues to caulk their ships, cracked and weakened by the long trips.

According to sixteenth-century chronicles, the very first oil export in history departed from Venezuela around 1539 when several barrels of *mene* were shipped to Spain. Rumors about the medicinal properties of the substance found around the Lake Maracaibo region prompted this shipment in an attempt to cure the gout that afflicted Emperor Charles V.

Far to the north in Canada, petroleum was first noted on record in 1719 when a Cree Indian named Wa-pa-su took a sample of oil sand to Hudson Bay's Henry Kelsey. However, petroleum wasn't drilled for until 1851 when Charles and Henry Tripp and their friend, Charles Nelson, dug an oil well in Woodstock—eight years before Drake succeeded in drilling a well in Pennsylvania. Although not commercial, it led to the formation of the first North American oil company, International Mining and Manufacturing (1851).

In the United States, the westward expansion promoted the search for oil. Settlers moving into the West to start a new life needed water during the drier summer months and salt for preserving food for the long winter months. So they dug wells for both water and brine. However, some of these crude hand-drilled wells also produced oil. Since the oil was a contaminant, it was considered a nuisance by the settlers, who either drained it into nearby creeks or poured it into sump pits where it was burned.

Samuel M. Kier, a Pennsylvanian who owned some brine wells, hoped to find some way to realize a profit from the "useless" petroleum byproduct. He bottled his oil, labeled it Pennsylvania Rock Oil, and advertised it far and wide for its healing properties. But the "snake oil" failed to find a ready market. Undaunted, Kier then devised a crude still to convert his petroleum into lamp oil used in lanterns. The material burned, but it had a bad odor and produced a heavy black smoke.

Although Kier was unsuccessful, he had opened the door toward new petroleum products. By his looking for a market for petroleum, the industry was born.

THE SEARCH FOR LIGHT

What was stirring up all of this new activity with petroleum and drilling? We need to recall what had happened a century earlier. Around 1750, the Industrial Revolution had rocked England to its foundation. Man's new knowledge and his ability to apply it resulted in forces that transformed the world completely. During the 100 years preceding the formation of the first North American oil company, Watt had improved the steam engine, America and France had undergone revolutions, Fulton had invented the steam boat, and Morse had developed the telegraph. These five events characterize the hundreds of inventions and happenings that were changing the shape of man's destiny. As more uses were found for oil, its need was clearly established.

By 1850, the idea of popular education had spread; more people could read and write, and popular newspapers and magazines published for the masses provided more reading material. In order to read, people needed light. Many still used tallow candles, which were expensive and ineffective, or sperm oil from whales, which was also expensive and not available to everyone.

Both in America and Europe, a search was on for a source of light that could be sold at a reasonable price, burn clearly, and provide effective illumination. In a few large cities, plants were built to process artificial gas from coal. Gas street lights and jets inside homes became popular, but the piping was expensive to install and the gas was only available in metropolitan areas. So the search—including attempts to produce a light source from petroleum—continued.

In 1849, James Young of Scotland obtained a patent for processing cannel coal, a bituminous carbonate containing much volatile material that burns brightly. He distilled it into what he called *coal oil*. It became popular almost immediately, and Young issued licenses on his patent for its production in the United States and Great Britain. A similar process was developed in Canada in 1854 by Dr. Abraham Gesner, who also licensed plants under an American patent. He called his product—which, like coal oil, was made from coal—*kerosene* from the Greek word *kēros*, meaning oil or wax. At that point, both coal oil and kerosene were the same coal-derived product.

The new liquid hydrocarbon caught on and quickly began replacing whale oil in lamps. By the end of the decade, over 50 plants were manufacturing kerosene from coal in the U.S. and, for a while, it looked as though coal oil were king. Meanwhile, however, others had been experimenting with producing kerosene from crude oil. Ultimately, coal gave way to crude, and the word "kerosene" came to be associated with oil, while "coal oil" was associated with coal.

With a better illuminating fluid, lamp makers perfected the kerosene or coal oil lamp, replacing the older, less efficient lantern. Col. A. C. Ferris, one of the major suppliers of lamp oil, was so excited about the superior lighting quality of kerosene that he immediately sent salesmen everywhere, buying crude oil for $20 per barrel. Much of Ferris' oil came from seeps, brine wells, or oil springs, but it was this sudden need for oil and the establishment of a firm price per barrel that brought about the next phase of the industry.

BIRTH OF AN INDUSTRY

Col. Edwin Drake had spent three years unsuccessfully trying to skim oil from the springs around Oil Creek, Pennsylvania. He had constructed a wooden boom across the creek and channeled the water through a series of lock-like skimmers that stopped the passage of the lighter oil and allowed the heavier water to flow on below. This process yielded about 1–3 barrels of oil per day—a very small quantity, considering the number of man-hours involved. So Drake decided to halt the project and formed the Seneca Oil Company to try again—this time using a new technique.

At this point, we need to take a step back in history to between 1807 and 1808. During these years, two brothers, David and Joseph Ruffner, had drilled a well for brine to a depth of 58 feet. Although they didn't find oil, this event represents a great moment in the development of the drilling industry. First, many of the fundamental techniques used later by Drake and his contemporaries were developed and used by the Ruffners. Second, the brothers used certain pieces of equipment in the foot-powered, spring-pole drilling rig that were early prototypes of modern drilling tools: a bit made from a $2\frac{1}{2}$-in. steel chisel, a drilling line, and—most importantly—a wooden casing. Prior to the Ruffners, there was no way to keep the earth from collapsing back into the hole and hindering progress. Finally, a way had been found to drill to greater depths without the sides of the hole caving in. All of the basic elements for drilling were now consolidated.

Fifty years later, Drake also assembled a primitive drilling outfit composed mainly of salt-well equipment (Fig. 1–1). It was a modification of the spring-pole technique: the cable-tool percussion drilling rig. Although the bit was still dropped into the earth, the works were powered by a steam engine, not men. After penetrating 30 feet of rock, Drake struck oil at 69 feet. That day, August 27, 1859, is noted as the birthday of the oil industry in the U.S. Even though James Miller Williams, an Ontario carriage maker, had completed the first commercially producing oil well one year earlier, Drake went one important step farther: he proved that oil could be obtained in sufficient quantities to

FIG. 1-1 Cable-tool rig like the one used by Drake.

meet the increasing demand by drilling through rock. At last the combination had come together: the need for oil, an established market and price, and a method of obtaining crude in quantity.

Almost overnight, Titusville and the Oil Creek area became the world's first oil boom town as buyers, would-be producers, and lessors swarmed to the "Cradle of the Oil Industry." However, men in many countries were seeking oil—Drake just happened to be the first to find it in quantity by drilling. He was far more fortunate in his selection of a site than anyone at the time realized.

By pure chance, the Drake well was located in an area where the pay sands were close to the earth's surface and could be easily reached with the primitive equipment at hand. In fact, the Oil Creek field was the shallowest and most productive ever discovered. This permitted large-scale production in à short period of time. The area was also easily accessible to the transportation facilities that were available. Thus, needed supplies could be brought in by existing railroads and the oil could be shipped out. As a final blessing, the oil was of high gravity (flowed easily) and was sweet (sulfur free). It could easily be refined into a high-quality kerosene with the unsophisticated processes then in use. Since it had a paraffin base, it could be made into lubricants. If Drake's oil had been sour crude (highly sulfuric), the pioneer refiners would not have known what to do with it.

NEW USES FOR PETROLEUM

Dozens of rigs soon covered the Pennsylvania landscape (Fig. 1–2). More and more needs and uses for petroleum and petroleum-based products were discovered. The new industry was still an infant when the Civil War burst across America, unleashing a technological age of machines that needed lubrication. After the war, rebuilding the nation and settling the West made the need for oil a demand. Primarily, two inventions boosted the appeal of oil: the incandescent light that led to oil-fired generating plants and the internal-combustion engine. Finally, there was a profit for those who were successful in finding new sources of oil with the new needs, demands, and applications for what had long been considered a waste byproduct of the oil industry: gasoline.

FIG. 1–2 Soon many towns began to look like this as hundreds of wells were drilled (Billie Linduff collection).

As man moved west, so did the oil rigs. A sleepy town of 75 residents could become a beehive of thousands of scurrying drillers and roustabouts within only 24 hours if oil were discovered nearby. The landscape was filled with rigs—sometimes so close that it was said a person could walk completely across town without touching the ground. In 1897, the Nellie Johnstone #1 became the first producing well in Indian Territory (near Bartlesville), for a time making the Osage Indian tribe the richest per capita nation on earth. But it was on January 10, 1901, on a marshy bit of wasteland near Beaumont, Texas, that the infant industry reached maturity.

SPINDLETOP

Spindletop—although that is not the name they originally chose—was the culmination of the dreams of two highly unalike but equally colorful oil men: Captain Anthony Lucas, a transplanted Slav, and the legendary Patillo "Bud" Higgins, a one-armed hellion turned Sunday school teacher.

As a youth, Higgins had lost an arm as the result of a prank. He and a group of companions had tossed a hornet's nest into a Baptist prayer meeting. A pursuing deputy sheriff fired a shot that struck young Higgins, the wound became infected, and the arm had to be amputated. The man was a carousing brawler, however, whose reputation spread until the townsfolk dreaded to see him come into town on Saturday nights.

But again fate stepped in. Higgins was passing another Baptist prayer meeting years later when he paused, stayed on to listen, and "got religion." One of the pillars of the church, George W. Carroll, was deeply impressed by Higgins' conversion and offered him a job in the real estate business. Higgins accepted and quickly became a successful and respected member of the community.

One day after a heavy rainstorm, Higgins happened to notice an unusual deposit of red clay that contrasted sharply with the black dirt of the area. He had a sample analyzed and found it was the type of clay from which bricks could be made. He had long been planning a new town, Gladys City, and he now had found a ready source of building materials. Sensing a certain market, Higgins had a group of investors join him in financing an experimental kiln. The business showed profit, but it was inefficient; so Higgins went north to study established kiln operations. In Pennsylvania and Ohio he learned the kilns were fired by gas and oil, more stable and efficient fuels than coal or wood.

In talking to oil men at the same time, he learned that many of the geological characteristics they were looking for in their search for oil were present in a small mound that rose from the marshes south of Beaumont. Higgins studied the area carefully and became convinced

FIG. 1–3 Early-day drill site. Note the boiler on the far left to furnish steam for power (Billie Linduff collection).

that the mound was sitting atop oil—much oil. He approached George Carroll with his theory, and both men formed the Gladys City Oil, Gas, and Manufacturing Company in 1892. For five years, using the crude water-well equipment that was still common, the company unsuccessfully sought oil. Finally, Higgins sold out and looked for another route.

As a last resort, he advertised for help in a trade journal. The man who answered the ad was Anthony Lucas. Lucas, a former mining engineer from Louisiana, had formed a theory that gas, oil, and sulfur accumulate around salt domes, common geographic phenomenon that often occur around the Gulf Coast. Since he had already visited Beaumont, he was familiar with what Higgins was talking about. Sensing an opportunity to test his theory, Lucas decided to join Higgins. The two were soon at work on yet another attempt to find oil at Gladys City—Higgins for his kilns and Lucas to test his hypothesis.

This time, however, Lucas decided to use steam-powered rotary drilling rigs. Rotary rigs were certainly not new; by 1901 more than 100 wells had been drilled in Texas alone using rotaries. But most of these devices were crude rigs powered by a mule walking in a circle, and rotaries had not gained wide acceptance in the industry. Lucas united all of the best elements of rotary drilling and created a better-working drilling rig.

FIG. 1-4 Spindletop—America's first gusher (courtesy Amoco Torch).

On January 10, 1901, the Lucas well had reached 1,020 feet. No one remembers who heard the noise first; it began with a low, menacing rumble that shook the earth. Then, as the crew fled, a geyser of mud shot into the sky, taking most of the derrick and rig with it. The noise ceased as suddenly as it began. The men timidly approached what was left of the rig and, with uncommonly good sense, shut down the boiler fires. No sooner had they done so than the rumblings began again. Once more the men ran as fast as they could as more mud came blowing out of the hole, only to be blown aside by a column of gas, immediately followed with a roar by a solid spout of green, heavy crude that reached 200 feet into the sky (Fig. 1–4).

For nine days the gusher continued to spout. (A wasteful practice, but most oil men believed a gusher was the only sign of a *true* discovery.) More than 800,000 barrels of oil formed a lake that spread nearly three-quarters of a mile. Finally it was capped by a collection of valves and fittings, the forerunner of today's Christmas tree. News of the discovery spread almost as quickly as the free-flowing oil, and within days Higgins' dreams of an ideal town, Gladys City, had turned into a maze of wooden derricks: Spindletop.

If Drake's well had given birth to an industry, Lucas and Higgins in turn gave birth to the age of liquid fuel. Spindletop represented a number of firsts: the first gusher in the U.S.,[*] the first proof of Lucas's theory about oil and gas forming under salt domes, the first large-scale success of rotary drilling equipment, and the first notable use of drilling mud. Now there was no looking back; the world would never be the same again.

DETHRONING KING COAL

Up until World War II, hydrocarbons were used extensively for all types of fuel. Coal was the primary source, constituting more than 80% of the needs of the United States for heating and supplying industrial power. However, a coal miner's strike during the period introduced a completely new face for the future.

Eastern coal miners led a strike in the late 1930s that crippled the nation. As it dwindled on and on, crude oil refiners had the opportunity to push petroleum and gas products. Industry, tired of haggling with the strikers, eventually began to convert from coal to oil and gas. World War II and postwar industry also added a new surge of petroleum activity. In a brief span of only 10 years, the pendulum had swung toward oil with 75% usage, coal with 19%, and other forms of energy with 6%.

This pattern has existed through today. Now, with our dependence on Middle Eastern oil becoming more of a deterrent toward petroleum use, Americans are again looking back to King Coal as a possible alternative energy source. However, it will be many years before any major changes will occur, and no one can second guess the future. For the next several decades—indeed, until the 21st century—we will all be using petroleum and its products for many of our needs.

[*]James, John, or Hugh Nixon Shaw (first name undocumented) brought in the first gusher in North America in 1856. The story goes that the Canadian bit through limestone at 200 feet near Oil Springs, Ontario. Instantly, he had a flowing well. The pressure was said to have thrown the tools out of the hole and driven oil high into the air. Like Spindletop, the roar of the gusher could be heard for miles around.

CHAPTER 2

The Search for Oil

The earth is 4–6 billion years old. When it was formed, it passed through a molten stage, then began to cool. During the cooling, the earth shrank and buckled to form a rough surface of igneous rock, which developed from the molten magma. As the atmosphere formed, rain fell. It struck the rough surface, collected, and flowed from the high spots into the low places to form the first rivers and oceans. During various geological eras, mountains were worn away through erosion and waste material was transported by rivers to the oceans.

Our barren earth continued this process for perhaps the first half of its life. At some point in time, life began in the oceans. The remains of countless tiny sea animals settled in the sediment of silt, layer upon layer. This marked the Cambrian period, about 550 million years ago. By the Devonian period, some 350 million years ago, vegetation and, later, animal life spread over the land masses (Fig. 2–1). The remains of these flora and fauna were eventually washed to the seas along with the silt brought down by rivers and streams. This mass of living matter contained hydrogen and carbon, the basic elements of crude oil and natural gas. Through geologic ages, the sands, mud, silts, and organic materials were transformed into layers of sedimentary rock.

SEDIMENTARY ROCKS

Of the three rock types, sedimentary rocks are the most important in terms of petroleum geology since deposits of oil and gas are most frequently found in them. Seldom do igneous and metamorphic rocks contain oil or gas.

As layer upon layer of sediment and animal/vegetation deposits were buried, they were compressed by the weight of the layers above (Fig. 2–2). Pressure, heat, and other factors—chemical, bacterial, and radioactive—changed the organic material into today's natural gas and oil.

Sedimentary rocks may occur as loose mud and sand or be hard and compacted, depending on how much pressure has been exerted and how old the layers are. The rocks are made up of clastic material, chemical precipitates, and organic/biorganic debris. The *clastic fragments* are

FIG. 2-1 How life has developed on earth during the various eras.

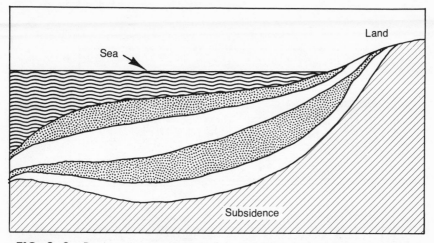

FIG. 2–2 During the sedimentation process, the weight of unconsolidated sediments compacts lower layers of sediments into sedimentary rock.

made up of broken and worn particles of other minerals, rocks, and shells that have been moved into place by erosion. *Chemical precipitates* were formed in place by the action of dissolved salts or the evaporation of water from entrapped sea water. *Organic/biorganic* debris is composed of shells and plant remains that accumulated in one place, such as a coral reef or peat bog (Fig. 2–3).

By understanding the composition and nature of sedimentary rock, petroleum geologists are able to determine where certain deposits may occur. These deposits are classified into three environments: *continental,* deposited on land by the wind; *transitional,* deposited at a delta formed at the mouth of a river (or between two such deltas); and *marine,* deposited in the ocean. Although sedimentation can occur anywhere on earth, a large part of the sedimentary rocks in the geologic column were probably deposited in transitional or marine environments in relatively shallow water on the continental shelf.

GEOGRAPHICAL MOVEMENT

If erosion had continued without compensation, the earth would have been completely leveled by now—one vast plain beneath the ocean. However, towering mountains rise high above sea level today. Some movement must have taken place to account for this. Even in Alaska, China, or Rumania, no one has ever witnessed an earthquake that resulted in more than a few feet of movement, yet seashells recovered from some of the world's deepest oil wells and from some of the

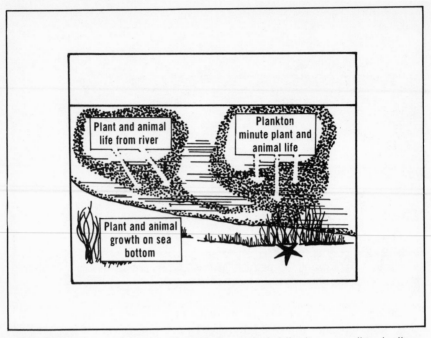

FIG. 2–3 Some sources of organic material that, according to the organic theory, provided the raw material for petroleum.

highest mountain walls show that ancient rocks have been raised or lowered thousands of feet during the course of time through repeated movements of the earth's crust.

Sedimentary rock is deposited in layers called *strata*, which are not strong enough to resist the earth's movement. Each time such a movement occurs, the strata is deformed. One common kind of deformation is the buckling of the layers into folds, as in both ancient and modern mountain chains. They may be small wrinkles or great troughs and arches many miles across.

Anticline is the name given to the arches or upfolds, and *synclines* are the troughs or downfolds (Fig. 2–4). These take many forms: they may be symmetrical with similar flanks or asymmetrical with one flank steeper than the other. The ends of both usually plunge sharply downward. A sharp anticline that is severely curved at the top is called a *dome*. Many domes contain a salt plug or core thrusting up in their center.

As the earth moves, almost all rocks become fractured. The cracks that are thus formed are called *joints*. If the rock on each side of the fracture shifts its position, a *fault* results (Fig. 2–5). The actual displacement on a fault may be either a few inches or up to several miles, such as the famous San Andreas fault located in California. The four main types

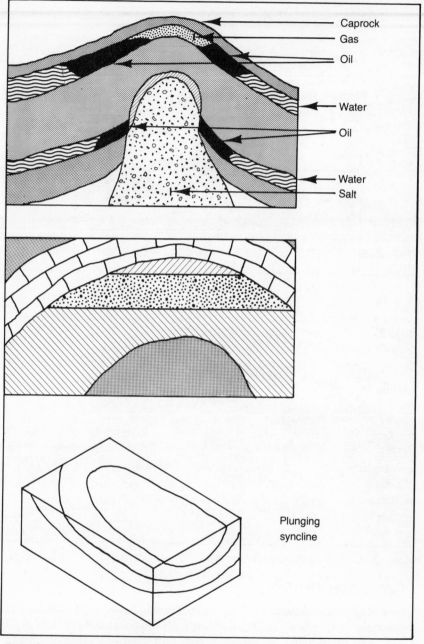

FIG. 2-4 A salt dome, an anticline, and a syncline.

of faults are normal, reverse, thrust, and lateral. Normal and reverse faults have vertical movement; thrust and lateral faults move mainly horizontally.

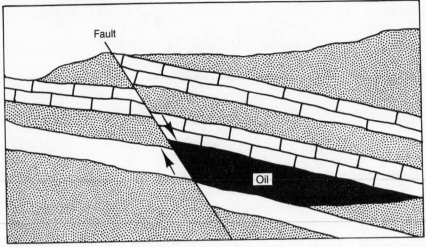

FIG. 2–5 A fault is a crack in the earth along which layers move.

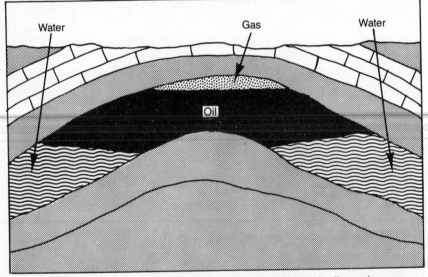

FIG. 2–6 Oil, gas, and water tend to separate into three layers.

ENTRAPMENT

Recall that sediment, tiny creatures, plant life, and small amounts of sea water were spread, layer upon layer, on the ocean floor. Heat, pressure, and the passing of countless ages turned these remains of living organisms into petroleum.

Oil and gas are usually not found where they were formed. Source rocks, in which the original tissue from the living organisms was trapped, are fine-grained and relatively impervious. They rarely hold oil and gas in anything but small quantities. Instead, the oil and gas move from the source rock upward toward the surface. Some escapes through faults to the surface. The gas disperses, the lighter oil evaporates, and a tar-like deposit of bitumen or pitch (the material Noah used to waterproof the ark) is left. Large quantities of the oil and gas never reach the surface, however. They migrate upward until they reach an impermeable barrier or cap rock and accumulate in place to form a *reservoir*. This barrier and the resulting reservoir are called a *trap*.

The upward movement is also accompanied by a separation of the oil, gas, and water (Fig. 2–6). The oil and gas rise as they displace the sea water that originally filled the pore spaces of the sedimentary rock. As they reach the impenetrable barrier, the materials separate. If you could place equal parts of natural gas, crude oil, and salt water in a sealed glass container, you would note that they eventually separate into three layers—the gas at the top, the oil in the middle, and the salt water at the bottom. This same separation occurs in the reservoir: gas is found in the highest part, then oil, and salt water at the bottom.

Not all of the salt water is displaced from the pore spaces, however. Often they contain from 10 to more than 50 percent salt water in the gas and oil accumulation. This remaining water, called *connate water,* fills the smaller pores and coats the surfaces of the larger openings.

Two kinds of impenetrable barriers can halt the upward progress of the oil, gas, and water: structural traps and stratigraphic traps (Fig. 2–7). *Structural traps* are formed because of a deformation in the rock layer that contains the hydrocarbons. Two common examples are fault traps and anticlines. A fault trap occurs when the formations on either side of the fault have been moved into a position that prevents upward migration. An anticline is an upward fold in the layers of rock, much like an arch. Petroleum migrates into the highest part of the fold, and its escape is prevented by an overlying bed of impermeable rock.

Stratigraphic traps result when the reservoir bed is sealed by other beds or by a change in porosity or permeability (discussed in the next section) within the reservoir bed itself. Stratigraphic traps include truncation (a tilted layer of petroleum-bearing rock cut off by a horizontal, impermeable rock layer), pinch-out (a petroleum-bearing formation gradually cut off by an overlying layer), a porous layer surrounded by impermeable rock, and a change in the porosity and permeability in the reservoir.

PROPERTIES OF A RESERVOIR

Often, people envision an oil or gas reservoir or pool as a large puddle of liquid far beneath the earth, like a subterranean pond. In reality,

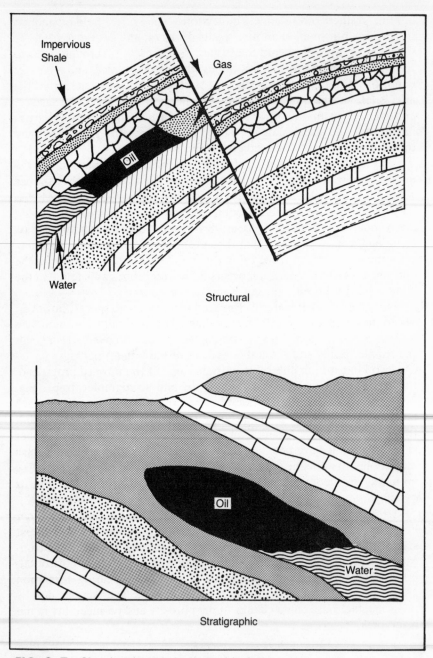

FIG. 2-7 Structural and stratigraphic traps.

though, the petroleum is entrapped in the pore spaces in solid rock. Before the reservoir can form, four qualifications must be met:

   1. A source bed must exist. This is the original layer that contained

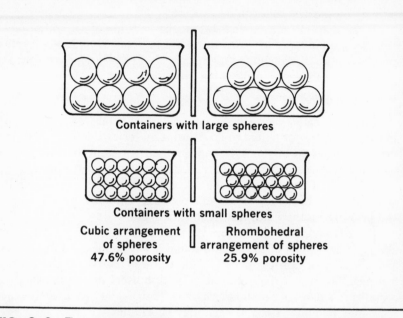

Containers with large spheres

Containers with small spheres

| Cubic arrangement of spheres 47.6% porosity | Rhombohedral arrangement of spheres 25.9% porosity |

FIG. 2–8 The arrangement and size of the spheres affects the porosity. Cubic arrangement can have a maximum porosity of 47.6%. A rhombohedral arrangement can yield a porosity of 25.9% (courtesy SPE-AIME).

the potentially petroleum-producing organisms and was submitted to the proper temperature and pressure.

2. The petroleum from the source rock must accumulate in a reservoir rock—a rock filled with holes and pores so the oil and gas can be collected (porosity).

3. The reservoir rock's pores must be interconnected so the oil or gas can move within the rock (permeability).

4. Some kind of closure or entrapment must exist that will prevent further upward movement of the fluids and will allow them to collect in one area.

If any one of these four points or characteristics is not present in an underground formation, a petroleum reservoir will not exist.

Porosity

Porosity, simply stated, is the capacity of the rock to hold fluids in its pores (Fig. 2–8). Mathematically, it is the volume of the nonsolid or fluid portion of the reservoir divided by the total volume. Thus, porosity is always expressed as a percentage. To visualize the concept, imagine a box full of balls of equal size stacked on top of each other so that only the most outward points touch the surrounding balls. The spaces in between the balls would be the pore spaces and would represent a porosity of 47.6%, the highest that can be expected.

If the balls are rearranged so that the balls are offset and fit down next to one another, the porosity would be reduced to 25.9%. Actual porosity in a reservoir may range from 3% to 40%, depending on the difference in the sizes of the grains and the way they join together.

Permeability

The permeability of a reservoir is the factor that determines how hard or easy it is for a fluid to flow through a formation (Fig. 2–9). It is not enough for the geologist to know that oil is present; he must also be able to determine how easy it will be for the oil to flow from the reservoir into the well. This is based on the viscosity or thickness of the fluid, the size and shape of the formation, the pressure, and the flow (the greater the pressure on the fluid, the greater the flow).

Permeability is usually expressed in units called *darcies,* after Henry d'Arcy, the French engineer who found a way to measure the relative permeability of porous rocks (1856). In most reservoirs, the average permeability is less than one darcy, so the usual figures are in thousandths of a darcy, or millidarcies (md). Permeability for a fine-grained sand may be 5 md; a coarse sand may be 475 md.

Formations of rock that hold concentrations of oil, gas, and water, the three kinds of reservoir fluids, are called reservoirs. The most common types are sandstones. Young sandstones are more porous and permeable than older types that have been buried more deeply and exposed to more pressure. Conglomerates, which are fairly common reservoir rock types, are consolidated gravel composed of pebbles of various sizes

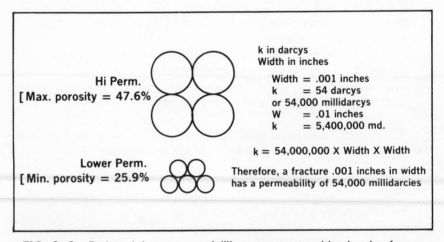

Hi Perm.
[Max. porosity = 47.6%

k in darcys
Width in inches

Width = .001 inches
k   = 54 darcys
or 54,000 millidarcys
W   = .01 inches
k   = 5,400,000 md.

k = 54,000,000 X Width X Width

Lower Perm.
[Min. porosity = 25.9%

Therefore, a fracture .001 inches in width has a permeability of 54,000 millidarcies

FIG. 2–9 Determining permeability measurement in darcies (courtesy Dresser Atlas).

held together by cementing materials such as clay or carbonate.

Many oil and gas reservoirs also occur in carbonate rocks, limestone, and dolomite. These rocks were formed by marine life that built a shell from calcium carbonate extracted from sea water. After death, the organic matter decayed and the skeletal material accumulated to form limestone. Some porous reef masses became reservoirs for petroleum.

Of the various types of sedimentary rocks, shales are the most common. They very rarely contain oil or gas in large quantities but, because they often contain finely scattered organic residues, shales may have been the most important source beds for hydrocarbons.

RESERVOIR FLUIDS

A fluid is any substance that will flow, including oil, water, and gas, the principal reservoir fluids. Oil and water are liquids; natural gas is a gas. But all three are fluids because they flow.

Water

Since most petroleum was initially formed from matter settling to the ocean floor, quite a bit of salt water remains in the rock pores. This water is called *connate water* because it remains from a time when the rock was first formed. Invariably, this water is dispersed throughout the reservoir, although it tends to accumulate more near the bottom of the pool. The separation is never complete, though; some water always remains with the oil and gas.

The water that occurs at the bottom of the reservoir is called, appropriately enough, *bottom water*. The water that collects at the perimeter of the reservoir is called *edge water*. As will be discussed, bottom water and edge water are important factors in getting the oil out of the ground and up the well.

Oil

Oil is lighter than water, so it tends to accumulate above the water layer. Some of the water stubbornly clings to the sides of the pores, though, and refuses to sink downward (Fig. 2–10). This water is called *wetting water* and is often present in the oil when the well is produced.

Gas

Natural gas is always associated with oil in a reservoir. Given proper conditions of pressure and temperature, the substance will stay in solution (dissolved) in the oil. When the temperature and pressure are lowered, the gas comes out of solution. Free gas (gas not in solution) tends to accumulate near the top of the reservoir.

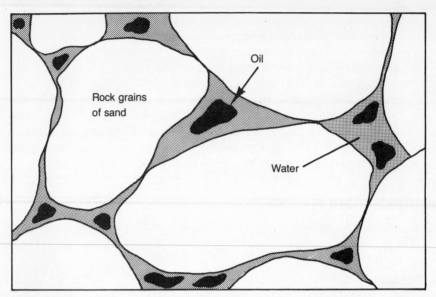

FIG. 2-10 Wetting water clings to the rocks, making them water wet.

FLUID FLOW

Three factors make it possible for the fluids in a well to move: pressure gradient, gravity, and capillary action. *Pressure gradient* is the difference between the pressures of two points. As in the weather, areas of high pressure tend to move toward areas of low pressure. If the pressure in the wellbore is less than the pressure in the formation, the fluids will flow toward the well bore and up the tubing. If the pressure isn't sufficient, a pump must be used to raise the fluids. *Gravity* aids a formation when a well is drilled to the lower portion. Gravity will push the fluids down toward the wellbore and eventually to the surface. *Capillary action* is the drawing or absorbing qualities of a material. A paper towel absorbs water by capillary action; oil and water act the same way in rock. They always try to move to areas with no fluids.

RESERVOIR DRIVES

Producing oil and gas—getting it out of the earth—takes energy. Nature usually supplies some of the energy. The fluids and gases are under pressure because of their depth, just as the pressure at the bottom of the ocean is greater than at the surface. The gas and water are two energy sources that help move the oil up the well.

Solution-gas Drive

A well drilled into a formation provides a means to relieve pressure (Fig. 2-11). If the formation contains oil and gas, the gas dissolved in the

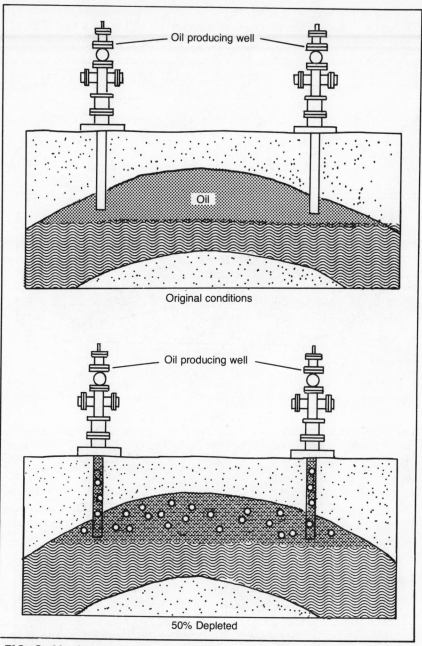

FIG. 2–11 Solution-gas drive reservoir.

oil forms bubbles, just like the bubbles that form in a bottle of soda when you take the cap off. As the pressure in the reservoir is reduced, the gas emerges and expands, driving the oil through the reservoir toward the wells and assisting in lifting it to the surface.

Wells drilled into a solution-gas drive accumulation usually must be pumped at first. The reservoir pressure graph will show a line of rapid and continuous decline. The gas-oil ratio will be low at first, then it will

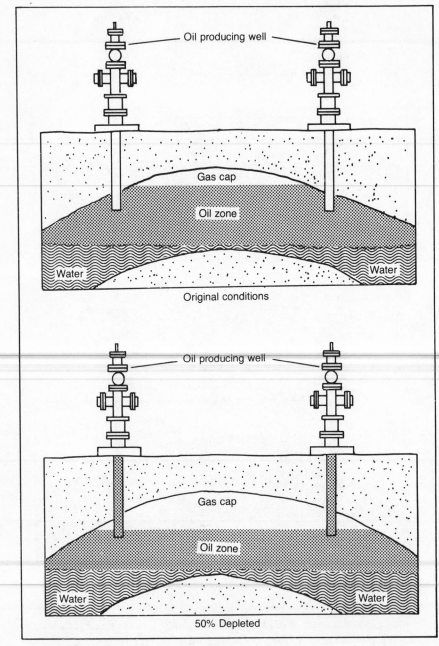

FIG. 2–12 Gas-cap drive reservoir.

rise to a peak and gradually begin to drop. How much of the original oil in the accumulation that can be recovered varies according to the property of the oil and the methods the producer uses. Generally, though, solution-gas or dissolved-gas drive is considered the least effective form of reservoir drive.

Gas-cap Drive

In many cases, there is more gas in a reservoir than the oil can retain in solution. This extra gas, since it is lighter than oil, rises to the top of the reservoir and forms a cap (Fig. 2–12). In gas-cap drive, the well is drilled into the petroleum-producing layer of the formation. As the oil begins to flow into the well bore, the pressure decreases and the expanding gas pushes down on the oil, forcing the oil up the well bore as the gas attempts to make its way up the well also.

Since more energy is present in this drive than in other depletion-type reservoirs, production is more stable and does not decline as rapidly. Depending on how large the original gas cap is, there will generally be a long flowing life. While the reservoir pressure will continue to decline, it will be at a slow rate and the gas-oil ratio will continue to rise. Because of the greater energy available for the drive, the recovery of original oil may be twice as much as that from a solution-gas drive reservoir.

Water Drive

When the formation containing an oil reservoir is fairly uniformly porous and continuous over a large area compared to the size of the oil reservoir itself, vast quantities of salt water exist in surrounding parts of the same formation, often directly in contact with the oil and gas reservoir. These tremendous quantities of salt water are under pressure and provide an additional store of energy to help produce the oil and gas (Fig. 2–13).

The expanding water moves into the regions of lowered pressure in the oil- and gas-saturated portions of the reservoir caused by the production of oil and gas and retards the decline in pressure. The expanding water also moves and displaces oil and gas in an upward direction from lower parts of the reservoir. By this means, the pore spaces vacated by oil and gas produced are filled with water. The oil and gas are progressively moved toward the well bore.

Generally speaking, the pressure will remain high just as long as the oil that is removed is replaced with an equal volume of water. As long as the pressure remains high, the gas-oil ratio will be low since there will be almost no free gas in the reservoir. If the high pressure is maintained, the well will continue to flow until the oil is exhausted and the well finally produces only water. Since water is more efficient than gas in displacing oil, recovery rates are considerably higher.

EXPLORATION

In the early days of the oil business, no methods had yet been devised for locating reservoirs except to search for seeps. Natural seeps near a

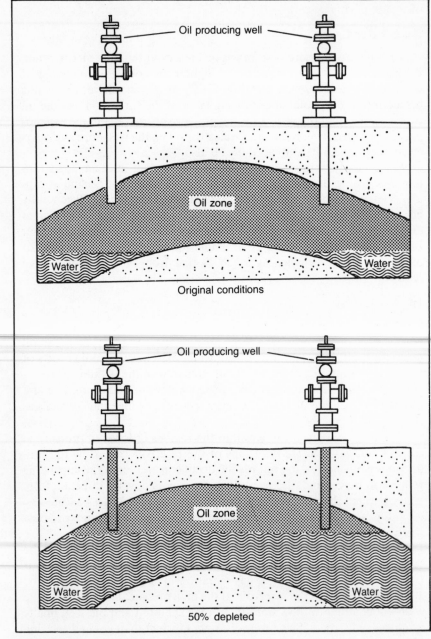

FIG. 2–13 Water-drive reservoir.

creek, as in Oil Creek, Pennsylvania, would leave an oily skim on the water, which drillers would recognize as a prospective location of oil. The same went for tar pits—oil was probably underneath.

It wasn't until 1885 that I. C. White came up with the theory that oil and gas accumulate along anticlines. From this point, the sciences of geology and geophysics began to play an increasing role in the oil industry.

The search for oil begins with geologists and geophysicists, using their knowledge of the earth to locate geographic areas that are likely to contain reservoir rock: areas with a source rock, a trap, porosity, and permeability. Once such an area is found, more specific tests and investigations are made in order to form some kind of picture of how the strata lie beneath the surface.

However, just looking for domes, seeps, and anticlines wasn't enough to locate an oil field. New geophysical methods were needed to give the explorationists an idea of what lay beneath the crust.

Gravimeter

One of the cheapest ways of locating a possible oil field is with a gravimeter. This instrument picks up a reflection of the density of the underlying rock. Salt is an important component in exploration, particularly around the Texas-Louisiana Gulf Coast. Quite often, huge columns or domes of salt rise up through the layers of strata like putty. When they finally near the surface, they raise the earth above to form a small hill. As the salt pushes through the underlying layers of strata, it forms an anticline—and anticlines are a potential reservoir location. Since salt is less dense than the surrounding rocks, its presence can be measured with the gravimeter. Where there is salt, there may be oil.

Magnetic Survey

Another valuable instrument that the geologist and geophysicist use is the *air magnetometer*. Magnetism is affected by the kind and depth of rock. Using the magnetometer, a magnetic survey of the area can be taken. The instrument is attached to an airplane and is flown over the area to be explored. The magnetometer measures the magnetic pull of the underlying layers of rocks to identify basement (usually Precambrian) rocks. These rocks contain large concentrations of a substance called *magnetite*. Therefore, a reading taken by the magnetometer will calculate how deep the basement rock lies. Then it is easy to determine the depth of the sedimentary basin lying above.

Seismic Survey

The *seismograph,* or *seismometer,* was first invented to detect shockwaves from earthquakes. However, geologists and geophysicists soon recognized its value in determining underground rock strata.

The theory behind the seismic survey is that subsurface structures can be deduced by measuring the transit times of sound waves generated by an explosion. First, the geophysicist decides on an area to be surveyed. He plans a pattern of dots that will spread out over an area of perhaps as much as a square mile or so (Fig. 2–14). Where each of these dots are marked on the map, a corresponding *geophone* is placed on the ground. A geophone is a circular, flat device that picks up seismic waves from underground. Each geophone is then attached by electric cables to recording instruments in a seismograph truck.

At a prearranged signal, either a charge of dynamite or nitroglycerin or the pounding of a vibrator truck begins. (The vibrator truck lowers a disc onto the earth and pounds the ground, alleviating the dangers and environmental disturbance of explosives.) As the waves from the explosion or the vibrator truck travel downward, they are reflected from the various layers below back to the geophones on the surface. These explosions or thumps continue at various points within the area until the entire location has been covered.

All this time, the messages are transferred back to the computers on the seismographic truck where the transit time for each layer is measured. The time it takes for waves to reflect back to the surface describes the thickness of the layer and the way it lies beneath the earth's surface.

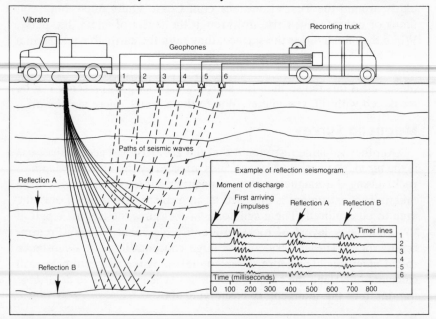

FIG. 2–14 When using seismic reflection surveying, the vibrator at point 1 creates shock waves that are reflected by the subsurface formations, picked up by the geophones, and recorded in the recording truck.

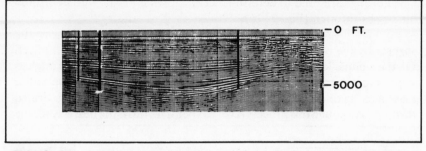

FIG. 2–15 A seismic section shows bright spots (courtesy Shell).

Through interpretation of these readings, exploration geologists deduce the shape and extent of the subsurface formation. Since traps often have a characteristic configuration, readings give a fairly accurate answer for the existence of oil.

Some scientists are now theorizing that bright spots (Fig. 2–15), which appear on the seismic sections, may be direct indicators of hydrocarbons. However, this is still a theory and has not been proven accurate in every case.

Remote Sensing

Another exploration technique is remote sensing. This involves using infrared (heat-sensitive) color photography or television to reveal ground water, salt-water intrusion, mineral deposits, faults, and other features not visible to the naked eye. The sensing equipment, carried aboard aircraft or satellites, is fed into special computers that compile the data and put it into a more usable form. In this way, the structures beneath the surface can be mapped.

STRATIGRAPHIC TECHNIQUES

In all of the methods that have been covered, the prime objective has been to form some kind of map of how the earth looks beneath our feet so the driller will have a better idea of where he should drill his well. So far, the techniques have allowed the geophysicist and geologist to map the features, but the earth itself hasn't been disturbed. There has been no physical, tangible evidence of exactly what lies below. If, in the explorationist's best judgment, the possibility of the existence of hydrocarbons is good, it's time to actually see what there really is down there.

Logs

Throughout the drilling operation, a log (or record) is kept of the various rock formations encountered, their type, and thickness. Several kinds of logs are used to determine characteristics of the formations the bit passes through (Fig. 2–16).

Sample logs. Pieces of rock cut by the drill bit and brought to the surface by the drilling fluid are retained and inspected for traces of oil. On the sample log, a record of the depth where the *cuttings* were made is kept, and any traces of hydrocarbons are noted (Fig. 2–16). The driller may also take *cores,* samples of the rock formation being drilled through. Although this process is very expensive, this is the best way for the

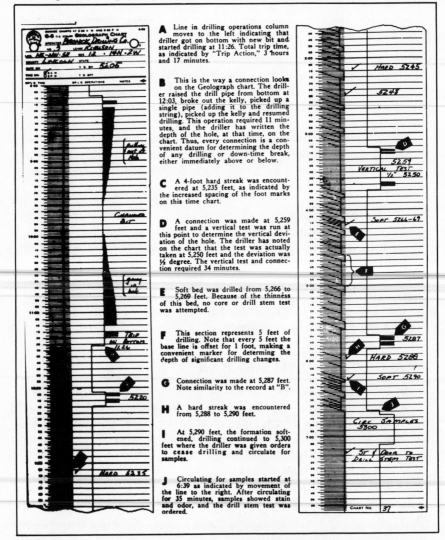

A Line in drilling operations column moves to the left indicating that driller got on bottom with new bit and started drilling at 11:26. Total trip time, as indicated by "Trip Action," 3 hours and 17 minutes.

B This is the way a connection looks on the Geolograph chart. The driller raised the drill pipe from bottom at 12:03, broke out the kelly, picked up a single pipe (adding it to the drilling string), picked up the kelly and resumed drilling. This operation required 11 minutes, and the driller has written the depth of the hole, at that time, on the chart. Thus, every connection is a convenient datum for determining the depth of any drilling or down-time break, either immediately above or below.

C A 4-foot hard streak was encountered at 5,235 feet, as indicated by the increased spacing of the foot marks on this time chart.

D A connection was made at 5,259 feet and a vertical test was run at this point to determine the vertical deviation of the hole. The driller has noted on the chart that the test was actually taken at 5,250 feet and the deviation was ½ degree. The vertical test and connection required 34 minutes.

E Soft bed was drilled from 5,266 to 5,269 feet. Because of the thinness of this bed, no core or drill stem test was attempted.

F This section represents 5 feet of drilling. Note that every 5 feet the base line is offset for 1 foot, making a convenient marker for determing the depth of significant drilling changes.

G Connection was made at 5,287 feet. Note similarity to the record at "B".

H A hard streak was encountered from 5,288 to 5,290 feet.

I At 5,290 feet, the formation softened, drilling continued to 5,300 feet where the driller was given orders to cease drilling and circulate for samples.

J Circulating for samples started at 6:39 as indicated by movement of the line to the right. After circulating for 35 minutes, samples showed stain and odor, and the drill stem test was ordered.

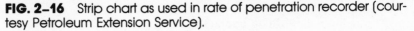

FIG. 2–16 Strip chart as used in rate of penetration recorder (courtesy Petroleum Extension Service).

explorationist to see exactly what is downhole. With a core, a paleontologist (scientist who studies fossils) can evaluate the fossils and decide whether that geologic period is known for hydrocarbon formation. As with the cuttings, the cores are marked alongside the depth on the driller's log.

Driller's logs. The driller's log is important because it measures the time to drill a certain distance as well as depth. If it takes a long time to dig through a particular strata, the formation is hard. If it takes only a few minutes to drill through several feet of rock, the formation is considered soft. This information helps the exploration geologist determine the kinds of formations and their possible porosity and permeability.

Electric logs. Electric logging is probably the most common of several wireline measurement methods used to assess the potential of rock formations (Fig. 2–17). After the hole has been drilled but sometimes at various intervals during drilling, instruments are lowered into the well on an electric cable (wireline). They send information to the surface about the type of formation, fluid content, and composition of the surrounding rocks by measuring the natural and induced electricity in formations. Since electricity is conducted through fluids better than through solids, formations that respond well to the test may contain hydrocarbon fluids.

Radioactivity logs. Radioactivity logs measure and record the effects of natural and induced radiation on the formation. Like the electric log, a tool is lowered into the well bore and is slowly drawn out. Along its path, it sends information to the surface that is computer-recorded on a record that resembles a graph or an electrocardiogram. These records can be studied and interpreted by an experienced geologist or engineer to see whether oil or gas exist and in what quantities.

Acoustic logs. Some instruments ping with sound waves and measure the reaction of the rock. The speed with which the wave moves through the rock depends on the composition of the rock and the fluid it contains. By calculating the time it takes for the sound to travel through a given formation, analysts can determine whether or not hydrocarbons may exist.

Maps

Once all of the data is collected and correlated, the geologist can help design subsurface maps of the underlying layers of rock. *Contour maps* of an area show the geologic structure relative to reference points called correlation markers. Contour lines are drawn at regular intervals of depth to add a third dimension. Increasing depth may also be illustrated with different colors.

Isopach maps illustrate variations in thickness between the correlation markers (Fig. 2–18). Again, color shading may be used for clarify

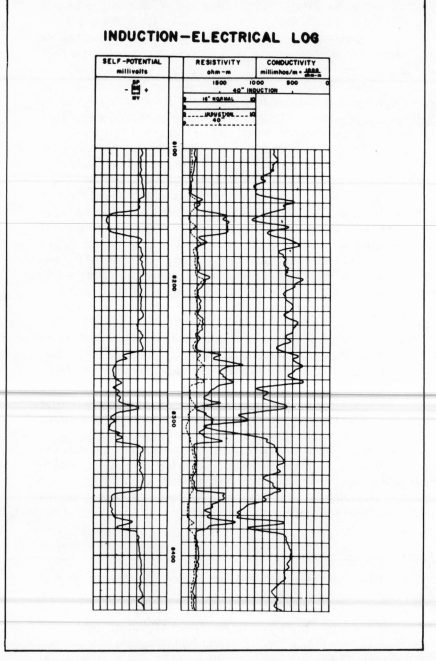

FIG. 2–17 Induction-electric log (courtesy Schlumberger).

and contrast. These maps are widely used in calculating reserves, planning secondary-recovery projects, and general exploration work.

Other maps can show faults and their intersections with other beds, porosity and permeability, variations in rock characteristics, and structural arrangement. However, all of these maps still only show a top view of the structure. For a complete picture, they must be supplemented with a *cross-sectional map* showing the side view (Fig. 2–19).

Conceptual Model

Once all of this data is collected and maps have been drawn of the formations, a conceptual model can be constructed. This is an idea of what the area looks like, how it is structured, and where accumulations of hydrocarbons are most likely to occur. With this information in hand, the oil company is finally ready to begin preparations for drilling.

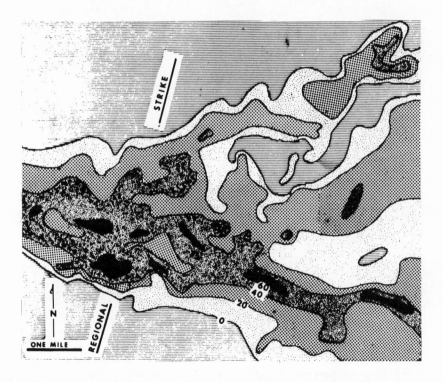

| None | 0-20 | 20-40 | 40-60 | Above 60 |

FIG. 2–18 Isopach of a total/interval (courtesy Petroleum Extension Service).

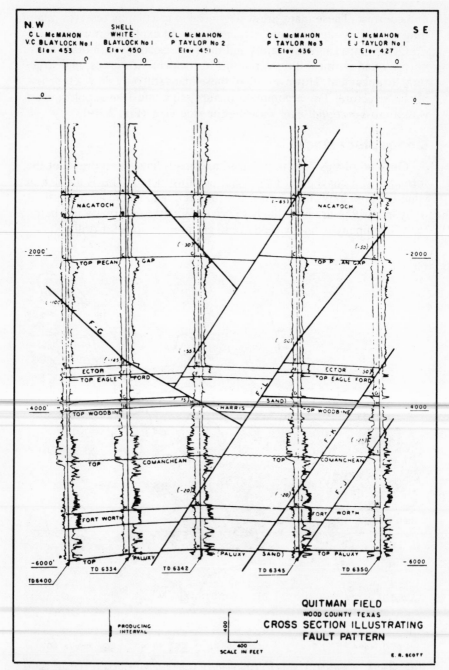

FIG. 2-19 Cross section of a fault pattern (courtesy Petroleum Extension Service).

CHAPTER 3

Leasing

Once a potential field has been found and the appropriate monies have been raised for the drilling venture, permission must be obtained from the landowner to enter his land and extract the oil and gas from the ground. In many parts of the world, the government owns all of the mineral rights. In the United States, however, the surface owner can often own the subsurface rights of the land also.

Of, course, it is impossible for oil companies to own all of the land on which they conduct operations. In fact, most of the search for petroleum is done either on government land or on land owned by individuals. Therefore, the company must get written permission to use the land. This is done through the work of the *landman* by means of a legal instrument called a *lease*. The lease sets out in detail the rights and responsibilities of both the company and the landowner. Because of the complexity of the contracts and the highly competitive nature of the work, leasing has become a very crucial aspect of the oil business.

Kinds of Leases

The American system of land ownership is patterned after the English system of "fee simple," or fee ownership. According to that old common law, a person owns his entire surface acreage as well as all the land beneath that down to the core of the earth. In essence, the property line extends from the fence line vertically downward to the center of the earth.

The shortcomings of indiscriminant settlement, the overlapping of claims, and boundary litigations were well known to the founding fathers. When new territory opened up, it was recommended that all lands be divided up according to the rectangular survey system. This system is based upon the establishment of a principal meridian and a base line. The principal meridian runs in a true north and south direction; the base line runs east and west at a right angle to the meridian. These lines help establish where a particular tract of land lies.

The legal description of a tract of rural land based essentially on an area referred to as a congressional township. The congressional township is an area six miles square and contains thirty-six square miles or sections.

After the principal meridian and base lines are laid out, the surveyor comes back to the point of beginning and lays out lines north and south every six miles on either side of the meridian. He then lays out lines parallel to the base line. Each square in the resulting grid is a township. Once the township and range lines are laid out, each of the squares is futher divided in 36 sections (Fig. 3–1). Numbering always begins with 1 in the upper right-hand corner and proceeds in a serpentine fashion to the bottom right-hand corner of the township to number 36.

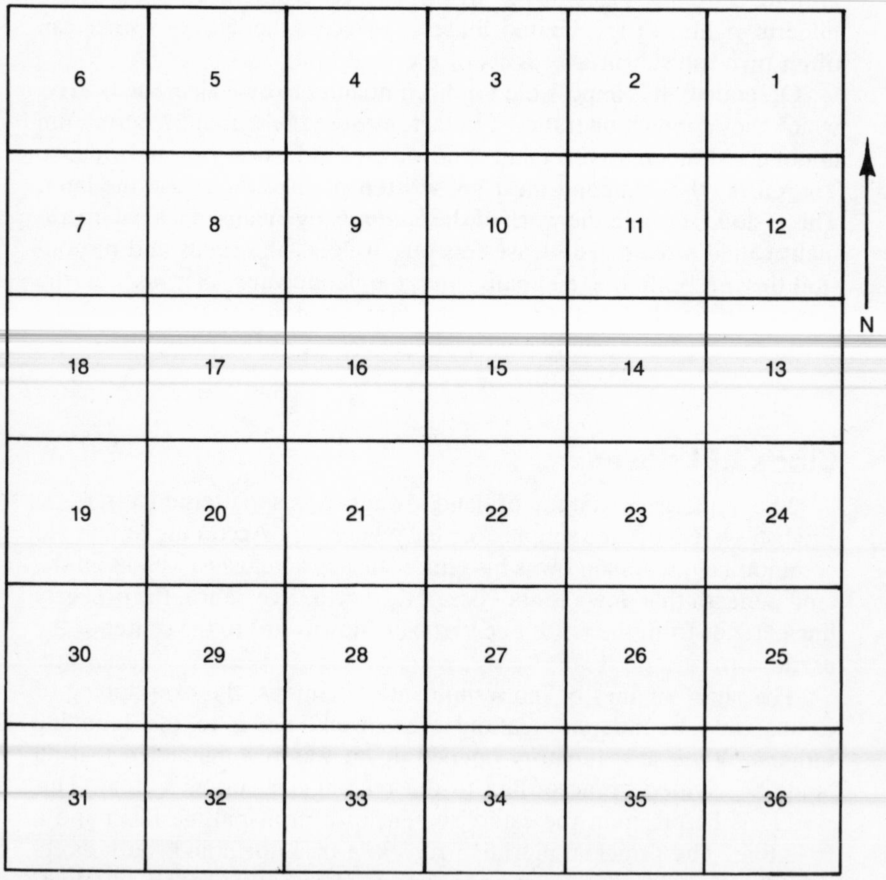

6	5	4	3	2	1
7	8	9	10	11	12
18	17	16	15	14	13
19	20	21	22	23	24
30	29	28	27	26	25
31	32	33	34	35	36

N →

FIG. 3–1 Each township is divided into 36 one-mile-square sections of land.

Under U.S. law, the landowner can search for and remove any minerals underneath his property or lease the land and those rights to another party. Today, in most states the land is divided into two estates or portions: the surface and the minerals. Sometimes one person owns both surface and mineral rights. Other times one person owns the surface and another person owns the minerals in the subsurface. If you live in a city, you're under this system. You own your house and the surface of the land, but the city owns all of the minerals under you.

When a person owns both mineral and surface rights to a piece of property, he may decide to transfer some of the rights to another party. If so, there are three different options he can exercise.

Lease Interest

A lease interest grants a lessee (an oil or gas company) certain subordinate rights and reserves other rights for the lessor (the landowner). However, the rights are not balanced. Although the company can come onto the landowner's property and extract minerals from the ground, it does not own any of the land. The company can usually come onto the land and prepare it for drilling, according to the terms of the lease. And it can remove oil or gas for the period of time specified in the lease, called the *primary term*. In exchange for granting these rights, the landowner retains certain provisions in the lease for his benefit:

1. He receives a bonus payment for granting the lease to the oil or gas company.
2. He receives delay rental payments if the drilling is held up for any reason or if the well is shut-in for a period of time.
3. He receives a royalty on the minerals that are extracted from beneath his land. This can either be paid in money or in minerals, according to the terms of the lease.
4. He retains the right to have ownership of the leased property and minerals revert back to him when the lease is terminated.
5. He may use the surface land in any way he wishes, providing it doesn't interfere with the lessor's operations that are outlined in the lease.

Mineral Interest

This kind of lease is often mistakenly called a royalty lease. In this option, the landowner may either sell the mineral rights to someone else or he may sell the land and retain the mineral rights. If he wishes to transfer the rights to the minerals, he signs a *mineral deed*. If he wishes to transfer only the monies received from the minerals, he signs a *royalty deed.*

Two major kinds of ownership states exist in the U.S., absolute ownership and nonownership. *Absolute ownership* means the landowner (the surface owner) owns all of the land that is not mined. The mineral owner owns the oil and gas and the subsoil it is mined from, but not the formation or surrounding rock. In other words, the surface owner could lease out mineral storage beneath the surface as long as it did not interfere with the lessor's operation.

In this estate, the surface owner must let the mineral owner have access to his minerals via the surface. This kind of system is used in Texas.

In Louisiana, a *nonownership* state exists. In this situation, the mineral owner does not own any of the subsoil that is mined. He only owns the oil and gas. If he brings up other minerals in the process, he cannot claim ownership.

In both systems, the mineral-rights owner has a separate estate from the surface-rights owner. If one estate should change hands or if the rights should be conveyed or sold, that does not affect the other estate. A mortgage can foreclose and the surface owner may have to move, but his successor still retains all of the rights from the lease. Likewise, one oil company might sell its lease rights to another oil company, but the terms from the lease would still be binding.

Royalty Interest

In a royalty interest, the owner reserves or deeds the royalty rights of the minerals to someone else. The major difference between royalty interest and mineral interest is that the owner of a pure royalty interest has no rights to enter the land and drill on it. The company cannot set up any operations or legally have any access to the minerals from the surface directly overhead. However, the lessor does not have to pay bonuses or delay rental charges, and he does not have to develop the property. He merely owns the proceeds of the minerals.

Government Leases

Quite of bit of the potential oil-producing land is owned by the federal and state governments. Some of this land can be leased to oil companies for exploration, drilling, and production. Leasing procedures are under the jurisdiction of the Department of the Interior (USDI) and the Bureau of Land Management (BLM). Offshore, procedures are set up by the Mineral Lands Leasing Act (1920), and Outer Continental Shelf (OCS) leasing is controlled by the OCS Lands Act.

Onshore Leasing

According to established policy, leasing of onshore oil and gas areas is governed by competitive and noncompetitive methods. For a known geological structure, leases of up to 640 acres and five years are given on a competitive basis. After enough interest is shown, bids are submitted to an agency and the highest wins the lease. Sometimes, *bonus bidding* is used. Here, sealed bids are submitted, based on a bid bonus with a fixed royalty and rental rate or on a bid royalty with a fixed rental rate and fixed bonus.

For noncompetitive land that is undeveloped, up to 2,560 acres and 10 years are granted. Sometimes this is done first come, first serve. In other cases, the tracts of land are chosen by lottery. More onshore lands are leased on a noncompetitive basis. Annual rental is usually far below competitive leases, and royalty fees can vary greatly.

Offshore Leasing

Periodically, the Dept. of Interior decides which tracts of its offshore (OCS) lands it should lease and it notifies the public that those areas are open for investigation. Seismic companies come in and conduct surveys and tests and an environmental impact hearing is held. Finally, the U. S. Geological Survey (USGS) places a minimum value on each tract and publishes the terms of the sale.

During offshore bidding, the tracts of land are announced and each sealed bid is read. The highest bid with the most attractive bonus may be selected the winner, or the government may reject all bids and retain the lease. If a lease is awarded, it must not be larger than 5,760 acres and is granted for five years. The minimum royalty rate for the government is 12.5%—often considerably higher. In addition, an annual rental is charged on each acre leased by the oil company.

Landowners

When the landman goes into the county courthouse to check on ownership of mineral rights, he needs to keep in mind certain rights and restrictions of landowners. The courts have had many battles throughout the years concerning the rights of landowners, but six major premises hold:

1. *Oil and gas are minerals and are part of the land.* Since they are part of the land, they are considered the property of the land-owner.

2. *When the minerals are mined, they become personal property.* Once a person extracts the oil or gas from the ground, the mineral owner makes these his personal property. He in essence owns the minerals as well as their value.

3. *The owner may either drill for the minerals or transfer his rights to someone else.* Although oil companies are usually selected to come onto a piece of property and produce the oil or gas, the owner does not have to go this route. Under law, he has every right to drill his own well and extract the minerals himself.

4. *The owner can drill a well and drain a reservoir even if it extends under his neighbor's property (rule of capture).* If the petroleum reservoir extends beyond the boundaries of an owner's property line, he may continue to drain the reservoir even if he is draining his neighbor's minerals. Since the minerals flow and there is no way to contain them within a boundary, it is up to the neighbor to build his own well and reap the profits from the minerals.

5. *The owner is liable for any damages he incurs to a common reservoir.* If, through poor drilling or production practices, the owner damages the reservoir because he depletes it too quickly, losing formation pressure, or he fractures the well so completely that the strata is damaged, he is liable and must compensate damages to the other landowners who share the reservoir.

6. *The owner is liable if he wastes oil and gas.* According to environmental protection guidelines, all oil and gas must be conserved. If an owner flares his natural gas or allows the crude to seep out over the land, he is responsible and must render reparations.

Once the landman understands all of these requisites for ownership, he must also consider the status of the owner he is dealing with. If the owner is a minor, the law will often not recognize any kind of legal binding contract made with him. The same goes for an incompetent. However, an illiterate can hold a valid lease as long as the proper witnesses and mark are in compliance with the law.

The landman must be careful of other types of arrangements, also. If the person living on the land is a life tenant without legal right to inheritance, the landman must deal with both that person and the remainderman—the person who inherits the rights. If both are not consulted, problems can arise later on. The same goes for owners of future interests, especially if they are not born yet.

Partnerships and groups of people can form yet another obstacle to obtaining a valid lease. A fiduciary, executor, or administrator of an estate may have the right to sell the land, but he may not have the right to grant a lease. For married people, much depends on the status of the state they live in. In some states, the wife has rights only on her dower property. In other states, the property is noncommunity property and both spouses must sign. With cotenants, all members should sign or problems may arise later on.

Lease Provisions

The sample lease in Fig. 3–2 is a good illustration of the kinds of clauses involved in a mineral lease. The numbered sections on the lease correspond to the appropriately numbered sections in the following discussion.

1. *Granting clause.* This paragraph sets out the company's rights and the mineral-owner's rights by naming the specific purpose of the venture and the privileges each party is awarded. In some states, it is automatically understood that the company has the right to enter and leave the property, the right to set up equipment, drill, and produce, and the right to reinject salt water into nonproductive formations.

2. *Habendum clause.* This sets out the initial term of the lease. Often, the first part of the clause will set up the time for exploration and drilling, and the second part will determine how long production may continue.

3. *Royalty clause* This is an agreement to pay the landowner a portion of the proceeds of the well, either in the mineral itself or in its equivalent cash value. The usual lease in the past has been one-eighth, but that figure is changing quickly with the recent drilling boom.

4. *Drilling and delay rental clause.* If the company cannot complete the well by the date specified in the habendum clause, it must pay the landowner a delayed rental fee or must terminate the lease. This protects the landowner from companies that might begin drilling and then abandon the well for a more productive field elsewhere.

5. *Pooling clause* This allows the oil company the option to pool or unite their leases with other leases in the area. The companies may share operating costs, and royalties can be shared on an equitable bases. This can provide more efficient recovery and less waste.

6. *Special provisions clause.* For the landowner, this clause states restrictions about drilling close to dwellings, burial stipulations, and payments for crop damage. For the oil company, it includes provisions for access to water, the right to move equipment on the land, and notice of change of ownership.

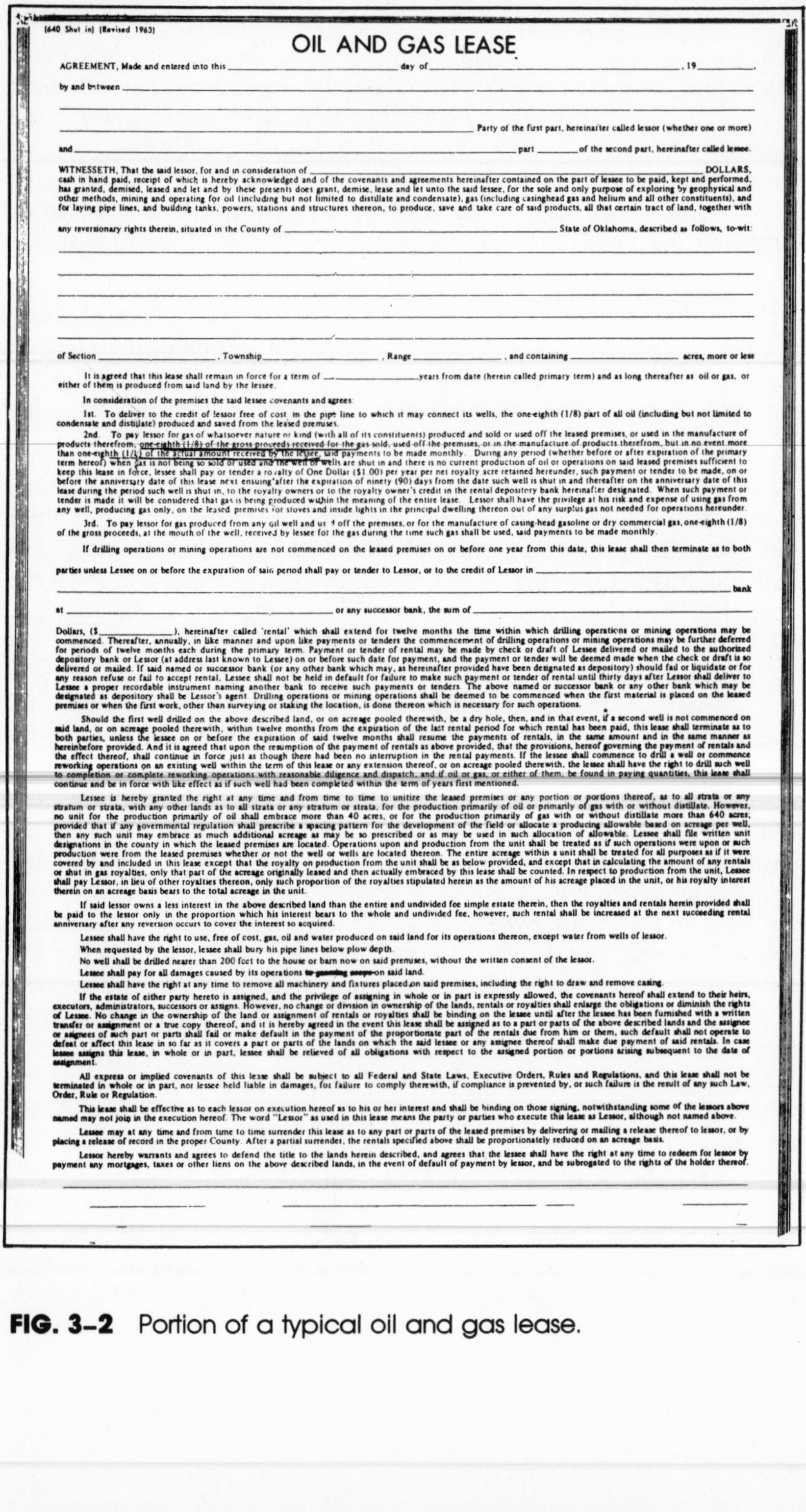

(640 Shut in) (Revised 1963)

OIL AND GAS LEASE

AGREEMENT, Made and entered into this ______________ day of ______________, 19______,

by and between ______________________________

____

______________________________ Party of the first part, hereinafter called lessor (whether one or more)

and ______________________________ part ________ of the second part, hereinafter called lessee.

WITNESSETH, That the said lessor, for and in consideration of ______________________________ DOLLARS, cash in hand paid, receipt of which is hereby acknowledged and of the covenants and agreements hereinafter contained on the part of lessee to be paid, kept and performed, has granted, demised, leased and let and by these presents does grant, demise, lease and let unto the said lessee, for the sole and only purpose of exploring by geophysical and other methods, mining and operating for oil (including but not limited to distillate and condensate), gas (including casinghead gas and helium and all other constituents), and for laying pipe lines, and building tanks, powers, stations and structures thereon, to produce, save and take care of said products, all that certain tract of land, together with

any reversionary rights therein, situated in the County of ______________ State of Oklahoma, described as follows, to-wit: 1

____

____

____

____

____

of Section ______________, Township ______________, Range ______________, and containing ______________ acres, more or less 2

It is agreed that this lease shall remain in force for a term of ______________ years from date (herein called primary term) and as long thereafter as oil or gas, or either of them is produced from said land by the lessee.

In consideration of the premises the said lessee covenants and agrees:

1st. To deliver to the credit of lessor free of cost in the pipe line to which it may connect its wells, the one-eighth (1/8) part of all oil (including but not limited to condensate and distillate) produced and saved from the leased premises.

2nd. To pay lessor for gas of whatsoever nature or kind (with all of its constituents) produced and sold or used off the leased premises, or used in the manufacture of products therefrom, one-eighth (1/8) of the gross proceeds received for the gas sold, used off the premises, or in the manufacture of products therefrom, but in no event more 3
than one-eighth (1/8) of the actual amount received by the lessee, said payments to be made monthly. During any period (whether before or after expiration of the primary term hereof) when gas is not being so sold or used and the well or wells are shut in and there is no current production of oil or operations on said leased premises sufficient to 4
keep this lease in force, lessee shall pay or tender a royalty of One Dollar ($1.00) per year per net royalty acre retained hereunder, such payment or tender to be made, on or before the anniversary date of this lease next ensuing after the expiration of ninety (90) days from the date such well is shut in and thereafter on the anniversary date of this lease during the period such well is shut in, to the royalty owners or to the royalty owner's credit in the rental depository bank hereinafter designated. When such payment or tender is made it will be considered that gas is being produced within the meaning of the entire lease. Lessor shall have the privilege at his risk and expense of using gas from any well, producing gas only, on the leased premises for stoves and inside lights in the principal dwelling thereon out of any surplus gas not needed for operations hereunder.

3rd. To pay lessor for gas produced from any oil well and used off the premises, or for the manufacture of casing-head gasoline or dry commercial gas, one-eighth (1/8) of the gross proceeds, at the mouth of the well, received by lessee for the gas during the time such gas shall be used, said payments to be made monthly.

If drilling operations or mining operations are not commenced on the leased premises on or before one year from this date, this lease shall then terminate as to both parties unless Lessee on or before the expiration of said period shall pay or tender to Lessor, or to the credit of Lessor in ______________________________ bank at ______________________________ or any successor bank, the sum of ______________________________

Dollars, ($______________), hereinafter called 'rental' which shall extend for twelve months the time within which drilling operations or mining operations may be commenced. Thereafter, annually, in like manner and upon like payments or tenders the commencement of drilling operations or mining operations may be further deferred for periods of twelve months each during the primary term. Payment or tender of rental may be made by check or draft of Lessee delivered or mailed to the authorized depository bank or Lessor (at address last known to Lessee) on or before such date for payment, and the payment or tender will be deemed made when the check or draft is so delivered or mailed. If said named or successor bank (or any other bank which may, as hereinafter provided have been designated as depository) should fail or liquidate or for any reason refuse or fail to accept rental, Lessee shall not be held in default for failure to make such payment or tender of rental until thirty days after Lessor shall deliver to Lessee a proper recordable instrument naming another bank to receive such payments or tenders. The above named or successor bank or any other bank which may be designated as depository shall be Lessor's agent. Drilling operations or mining operations shall be deemed to be commenced when the first material is placed on the leased premises or when the first work, other than surveying or staking the location, is done thereon which is necessary for such operations.

Should the first well drilled on the above described land, or on acreage pooled therewith, be a dry hole, then, and in that event, if a second well is not commenced on said land, or on acreage pooled therewith, within twelve months from the expiration of the last rental period for which rental has been paid, this lease shall terminate as to both parties, unless the lessee on or before the expiration of said twelve months shall resume the payments of rentals, in the same amount and in the same manner as hereinbefore provided. And it is agreed that upon the resumption of the payment of rentals as above provided, that the provisions, hereof governing the payment of rentals and the effect thereof, shall continue in force just as though there had been no interruption in the rental payments. If the lessee shall commence to drill a well or commence 5
reworking operations on an existing well within the term of this lease or any extension thereof, or on acreage pooled therewith, the lessee shall have the right to drill such well to completion or complete reworking operations with reasonable diligence and dispatch, and if oil or gas, or either of them, be found in paying quantities, this lease shall continue and be in force with like effect as if such well had been completed within the term of years first mentioned.

Lessee is hereby granted the right at any time and from time to time to unitize the leased premises or any portion or portions thereof, as to all strata or any stratum or strata, with any other lands as to all strata or any stratum or strata, for the production primarily of oil or primarily of gas with or without distillate. However, no unit for the production primarily of oil shall embrace more than 40 acres, or for the production primarily of gas with or without distillate more than 640 acres; provided that if any governmental regulation shall prescribe a spacing pattern for the development of the field or allocate a producing allowable based on acreage per well, then any such unit may embrace as much additional acreage as may be so prescribed or as may be used in such allocation of allowable. Lessee shall file written unit designations in the county in which the leased premises are located. Operations upon and production from the unit shall be treated as if such operations were upon or such 6
production were from the leased premises whether or not the well or wells are located thereon. The entire acreage within a unit shall be treated for all purposes as if it were covered by and included in this lease except that the royalty on production from the unit shall be as below provided, and except that in calculating the amount of any rentals or shut in gas royalties, only that part of the acreage originally leased and then actually embraced by this lease shall be counted. In respect to production from the unit, Lessee shall pay Lessor, in lieu of other royalties thereon, only such proportion of the royalties stipulated herein as the amount of his acreage placed in the unit, or his royalty interest therein on an acreage basis bears to the total acreage in the unit.

If said lessor owns a less interest in the above described land than the entire and undivided fee simple estate therein, then the royalties and rentals herein provided shall be paid to the lessor only in the proportion which his interest bears to the whole and undivided fee, however, such rental shall be increased at the next succeeding rental 7
anniversary after any reversion occurs to cover the interest so acquired.

Lessee shall have the right to use, free of cost, gas, oil and water produced on said land for its operations thereon, except water from wells of lessor.

When requested by the lessor, lessee shall bury his pipe lines below plow depth.

No well shall be drilled nearer than 200 feet to the house or barn now on said premises, without the written consent of the lessor.

Lessee shall pay for all damages caused by its operations ~~to growing crops~~ on said land.

Lessee shall have the right at any time to remove all machinery and fixtures placed on said premises, including the right to draw and remove casing.

If the estate of either party hereto is assigned, and the privilege of assigning in whole or in part is expressly allowed, the covenants hereof shall extend to their heirs, executors, administrators, successors or assigns. However, no change or division in ownership of the lands, rentals or royalties shall enlarge the obligations or diminish the rights of Lessee. No change in the ownership of the land or assignment of rentals or royalties shall be binding on the lessee until after the lessee has been furnished with a written transfer or assignment or a true copy thereof, and it is hereby agreed in the event this lease shall be assigned as to a part or parts of the above described lands and the assignee or asignees of such part or parts shall fail or make default in the payment of the proportionate part of the rentals due from him or them, such default shall not operate to defeat or affect this lease in so far as it covers a part or parts of the lands on which the said lessee or any assignee thereof shall make due payment of said rentals. In case lessee assigns this lease, in whole or in part, lessee shall be relieved of all obligations with respect to the assigned portion or portions arising subsequent to the date of assignment.

All express or implied covenants of this lease shall be subject to all Federal and State Laws, Executive Orders, Rules and Regulations, and this lease shall not be terminated in whole or in part, nor lessee held liable in damages, for failure to comply therewith, if compliance is prevented by, or such failure is the result of any such Law, Order, Rule or Regulation.

This lease shall be effective as to each lessor on execution hereof as to his or her interest and shall be binding on those signing, notwithstanding some of the lessors above named may not join in the execution hereof. The word "Lessor" as used in this lease means the party or parties who execute this lease as Lessor, although not named above.

Lessee may at any time and from time to time surrender this lease as to any part or parts of the leased premises by delivering or mailing a release thereof to lessor, or by placing a release of record in the proper County. After a partial surrender, the rentals specified above shall be proportionately reduced on an acreage basis.

Lessor hereby warrants and agrees to defend the title to the lands herein described, and agrees that the lessee shall have the right at any time to redeem for lessor by payment any mortgages, taxes or other liens on the above described lands, in the event of default of payment by lessor, and be subrogated to the rights of the holder thereof.

FIG. 3–2 Portion of a typical oil and gas lease.

7. *Obligations of producer clause.*   Once the oil company has assigned the lease, it must commit itself to explore for oil and produce the formation to the best of its ability. If it doesn't, this is terms for ending the contract, for the landowner isn't making the money he deserves on his minerals.

To release the lease (generally it is because of failure to comply with the terms) the lessee abandons its rights, the lessee surrenders its rights, or the lessor forfeits the property rights. Other conditions can arise, but these are the four most common.

CHAPTER 4

Drilling and Completion

Once a well site is selected and leased, there are still many factors to take into consideration before the actual drilling begins. If the well is located outside the United States, the operating company may elect to subcontract the drilling to an independent drilling company. There will probably be several other highly specialized subcontractors involved before the well is completed.

The end result the operating company is seeking is a producing well—a well of steel casing with cement, extending from the surface downward into the oil-bearing formation with the necessary production equipment in place, that has been drilled as rapidly and economically as possible.

To accomplish this, a great deal of planning and forethought must go into the drilling program (as the blueprint of a well is called). Since much of the drilling program is determined by the formation itself, planning for the well begins at the bottom of the hole, not at the surface.

If the site is in a shallow, low-pressure formation where water saturation could be increased by use of water-based drilling fluids with the result that permeability would be decreased, a cable-tool rig may be selected. If, on the other hand, high pressures may be encountered, then a rotary rig might best be used so that the drilling fluid mixture can be used to regulate the pressure. In most cases, rotaries drill more rapidly than cable-tool rigs, and shorter time means less payroll to meet for both workers and supervisors.

Determining whether to use a cable-tool or rotary rig depends on three factors: character of the producing formation, cost per foot, and nature of the formations that must be drilled through. Today, we have a choice between the two major kinds of systems. But in the early days of drilling, matters were a bit different.

44

The Earliest Wells

Man, for one reason or another, has been digging holes in the earth's surface since earliest known times. The first wells were mainly for water and irrigation. Hand-dug wells were usually bare holes dug down into the soil and rock, but occasionally they were lined with rock to prevent internal collapse.

As man's technology improved, so did the methods of digging. Crudely shaped tools were first used, and later digging implements of bronze and iron appeared. Debris was at first handed up out of the hole in a basket. Later, it was hauled out with the aid of crude ropes and a windlass.

No one knows who was the first civilization to drill instead of dig, but by 600 B.C. the Chinese were using percussion tools–the forerunners of cable tools—to dig brine wells. By 1500 A.D. they were drilling down to 2,000 feet. Their rigs were constructed almost completely from bamboo, except for the bit on the end.

One of the biggest inventions connected with drilling came from two Americans, David and Joseph Ruffner. They were trying to dig a brine well near Charleston, West Virginia, but the walls kept collapsing. To support the sides of the hole, they cut up sections of a hollow tree and pushed it down as the hole was dug deeper. A man down within the hollow log used a pickax to loosen the rock and cuttings were lifted out in a whiskey barrel that was cut in half. In this fashion, the first casing and bailer were invented.

FIG. 4–1 Spring-pole technique used by the Ruffner Brothers.

The Ruffners also incorporated another drilling technique that had been used before but which was fairly novel in the U.S. They obtained a metal quarry bit and attached it to a rope, which was in turn tied to the small end of a 20-foot pole (Fig. 4–1). The larger end of the pole was anchored to the ground and supported in the center by a forked stick with the bit positioned directly over the hole. Basically, it worked like a diving board—in fact, it was called the spring-pole technique. Workers would stand in loops on the free end of the stick and kick downward, causing the bit to strike bottom. When they released their weight, the bit would spring free. In this way, the well was eventually dug.

By today's standards, it seems hard to believe the early tools could reach any depth at all. The drill bit, which was forged on an anvil, had a convex chisel point about four inches long. The shank of the bit was about 12 inches long and was screwed into the sinker, a square iron bar about 20 feet long that weighted the bit. This in turn was attached by a threaded joint to the poles. It wasn't until Drake turned over his drilling effort to Uncle Billy Smith that drilling entered the next phase in its development.

Cable-Tool Drilling

Drake and Smith revolutionized the spring-pole concept when, in 1859, the rig at Oil Creek was modified to use steam power instead of men springing on a pole and to use a line instead of a rod. By 1890, what was to be known as the standard cable-tool rig had been developed (Fig 4–2). It consisted of a wooden derrick, steam power plant, wooden band wheel, bull wheel, calf wheel, and sand reel. Drilling cable was hemp rope wound on the bull wheel. Casing was run and pulled with the calf wheel using wire rope since extra strength was needed. Wire rope was also used on the sand line because of abrasion.

Bull wheel. In drilling, the bull wheel was used to spool the drilling rope, which had to be long enough to reach the well bottom. The bull wheel was also used to pull the string of tools from the hole. Power was provided by direct drive from the engine to the band wheel to the bull wheel. There was no clutch, so when the driller had to stop the wheel he threw the rope off by sticking a board between it and the pulley. Later, the bull rope was replaced by a chain drive so drillers no longer had to duck flying ropes.

Calf wheel. The drilling rope ran over a single pulley in the crown and it couldn't develop any more pull than it had at the bull wheel. When casing had to be run, there simply wasn't enough power available. Another spool was added, called the calf wheel, together with a multiple-sheave crown block and multiple-sheave traveling block. With this,

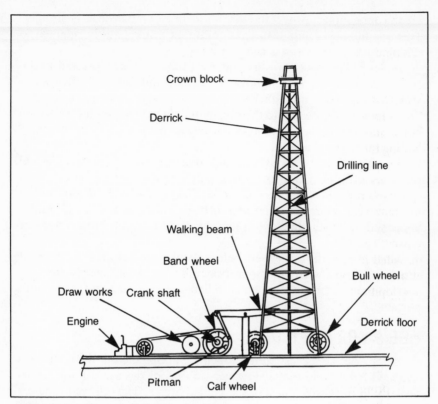

Crown block

Derrick

Drilling line

Walking beam

Band wheel

Bull wheel

Draw works   Crank shaft

Engine

Derrick floor

Pitman   Calf wheel

FIG. 4–2 Standard cable-tool rig.

more weight could be pulled. This was the forerunner of the draw works, crown block, and traveling block of today's rotary rigs.

Walking beam.   Many of the early rigs were permanent structures, so the walking beam that raised and lowered the drill bit later raised and lowered the sucker rods during production. The walking beam was driven by a crank and pitman. For drilling, it worked off the crank on the band wheel. Since there was no clutch, the driller pulled a cotter pin and slipped the pitman off the band-wheel crank to disconnect the beam to bail or to pull tools.

Tool joints and tongs.   To make and break the threaded connections between the parts of the tools, a circle jack was used. As the name implies, this was a semicircular toothed rack bolted to the rig floor. One end of a large wrench was connected to a pinion and then jacked along the rack, with a second wrench attached to the floor.

The tool joint also began on the early cable-tool rigs. The tool string included bit, sinker bars, and jars—all solid and connected with threaded joints. The drillers soon learned that they could make and break connections faster if the threads were tapered instead of straight. This was the beginning of the tool joint that would later be adapted to drillpipe when the rotaries came into use.

Bailer. The bailer was a cylindrical tube that was lowered in the hole with the sand line to remove excess fluid and cuttings. There was a metal loop at the top to attach the sand line to and a trap at the bottom. The trap would close due to the weight of fluid as it was pulled out of the hole. Once on the surface, the trap would be tripped and fluid would be dumped into the slush pit.

Steel and Safety. Two things lacked in the early standard cable-tool rig: steel and safety brakes. Often, only a board wedged between the spokes of the bull wheel and the derrick legs suspended the tools out of the hole. Hand brakes often kicked back when applied. If the driller happened to be standing over them, he could expect a broken bone or crushed skull.

Most of the derricks used before 1900 were wooden, permanent structures. The advent of temporary steel structures produced a superior derrick that became safer, more reliable, more powerful, and more portable.

Cable-Tool Rigs Today

After Lucas's success with a rotary rig at Spindletop, many oil men switched from cable tools. However, cable-tool rigs are still used, particularly in the eastern part of the United States. The only requisite for cable-tool drilling is that the well be shallow and that the formation be solid enough that there will be little damage. Although simple and slow, cable-tool drilling can be one of the cheapest methods of drilling.

Rotary Drilling Rigs

Like cable-tool rigs, the rotaries have gone through a period of evolution that has made them safer, more efficient, more powerful, and capable of drilling literally miles into the earth's crust. The concept of boring into the earth with a rotating motion is not new. As early as 3,000 B.C., the Egyptians were implementing methods still used in theory today.

During the 19th century, several patents were issued for rotary drilling systems. One of the earliest issued for a machine that used a rotating tool, hollow drill rods, and circulating fluid to remove the cuttings was granted to Robert Beard (1844). His device included a hydraulic swivel, a hollow square kelly, a belt-driven rotary machine, a splined quill and chuck, hollow drill rods, a fishtail bit, and circulating fluid to remove the cuttings. Thus, the stage was set for Spindletop where Lucas culminated

FIG. 4–3 Engine and draw works used at Spindletop (courtesy Petroleum Extension Service).

FIG. 4–4 Star steam-powered mobile rig.

all the knowledge and techniques of centuries to prove the value of rotaries in drilling in soft formations to great depths where cable tools could not be used (Fig 4–3). It ushered in a boom not just along the Gulf Coast area, but in East Texas, northern Louisiana, southern Arkansas, Mississippi, and out to California's San Joaquin Valley. Before long, rotary rigs were being exported around the world (Fig. 4–4).

Rotary Rigs Today

Today, 80–90% of all wells dug use the rotary drilling system (Fig. 4–5). Toothed bits cut up the rock, and complex mixtures of drilling fluids circulate cuttings out of the hole, eliminating the need for the time-consuming bailer. The main disadvantage of the system is that the drilling fluids mask the pay zones, unlike the cable-tool method. Since the fluid forces formation fluids back, a potential pay zone might be passed up because no shows of petroleum could enter the well bore.

Drilling Contracts

Before a contract can be drawn up between the operator and the drilling company, several factors must be determined. The operator must decide which kind of drilling method he will use, the proper rig size, cost of the rig, casing requirements, bit program, speed, and the

FIG. 4–5 A modern rotary rig (courtesy Noble Drilling Co.)

Offshore International Contract
Revised June, 1975
0.5M 9-76

International Association of Drilling Contractors
INTERNATIONAL DAYWORK DRILLING CONTRACT — OFFSHORE

THIS AGREEMENT, dated the _____ day of _____ , 19 _____ , is made between:

_____ , a corporation organized under the laws of

_____ , located at _____ ,
and hereinafter called Operator,

and: _____ , a corporation organized under the laws of

_____ , located at _____ ,
and hereinafter called Contractor.

WHEREAS, Operator desires to have offshore wells drilled in the Operating Area and to have performed or carried out all auxiliary operations and services as detailed in the Appendices hereto or as Operator may require; and

WHEREAS, Contractor is willing to furnish the drilling vessel complete with drilling and other equipment, (hereinafter called the "Drilling Unit"), insurances and personnel, all as detailed in the Appendices hereto for the purpose of drilling the said wells and performing the said auxiliary operations and services for Operator.

NOW THEREFORE THIS AGREEMENT WITNESSETH that in consideration of the covenants herein it is agreed as follows:

ARTICLE I — INTERPRETATION

101. Definitions

In this Contract, unless the context otherwise requires:

(a) "Commencement Date" means the point in time that the Drilling Unit arrives at the place in or near the Operating Area designated by Operator, or at a mutually agreeable place in or near the Operating Area, or on arrival at the first drilling location, whichever event occurs earliest;

(b) "Operator's Items" mean the equipment, material and services which are listed in the Appendices that are to be provided by or at expense of Operator;

(c) "Contractor's Items" mean the equipment, material and services which are listed in the Appendices that are to be provided by or at expense of Contractor;

(d) "Contractor's Personnel" means the personnel to be provided by Contractor from time to time to conduct operations hereunder as listed in the Appendices;

(e) "Operating Area" means those areas of the seabed and subsoil beneath the waters offshore _____ in which Operator may from time to time be entitled to conduct drilling operations;

(f) "Operations Base" means the place or places on shore designated as such by Operator from time to time.

(g) "Affiliated Company" means a company owning 50% or more of the stock of Operator or Contractor, a company in which Operator or Contractor own 50% or more of its stock, or a company 50% or more of whose stock is owned by the same company that owns 50% or more of the stock of Operator or Contractor.

102. Currency

In this Contract, all amounts expressed in dollars are United States dollar amounts.

103. Conflicts

The Appendices hereto are incorporated herein by reference. If any provision of the Appendices conflicts with a provision in the body hereof, the latter shall prevail.

104. Headings

The paragraph headings shall not be considered in interpreting the text of this Contract.

105. Further Assurances

Each party shall perform the acts and execute and deliver the documents and give the assurances necessary to give effect to the provisions of this Contract.

106. Contractor's Status

Contractor in performing its obligations hereunder shall be an independent contractor.

107. Governing Law

This Contract shall be construed and the relations between the parties determined in accordance with the law of _____ , not including, however, any of its conflicts of law rules which would direct or refer to the laws of another jurisdiction.

ARTICLE II — TERM

201. Effective Date

The parties shall be bound by this Contract when each of them has executed it.

202. Duration

This Contract shall terminate:

(a) immediately if the Drilling Unit becomes an actua' ¬structive total loss;

_____ months after ¬¬ ¬f termination from Operator but Operator may not give

¬ until ¬¬¬ ¬ Date, (or ¬¬ ¬ ¬en being conducted on

FIG. 4–6 Typical IADC daywork contract.

Published by
Division of Production
American Petroleum Institute
300 Corrigan Tower
Dallas 1, Texas

(AP)

Model Form 4A2
First Edition
August, 1962

EXHIBIT A
BID SHEET AND WELL SPECIFICATIONS

To:_____

Gentlemen:

We solicit your bid to drill and complete the hereinafter designated well.

This bid form has been filled in by us to the extent necessary to disclose the manner in which we desire the well to be drilled. If you desire to submit a bid, please complete this instrument in every respect, execute the original and two copies, and return to our office at_____not later than_____a.m. p.m.
_____, 19____

Very truly yours,

_____
Operator

By:_____

1. NAME AND LOCATION OF WELL:

Well Name
and Number _____ Parish
County_____ State_____
Field Name_____ Well location and land description_____

2. COMMENCEMENT DATE:

Contractor agrees to commence actual_____operations at the above location on or before_____,
19_____, or, in the event Operator is to clear and grade location and furnish roadway or other ingress or egress facilities, within_____days from the date of completion of the clearing and grading and construction of roadway, or such other ingress or egress facilities, whichever is the latter.

3. DEPTH:

Subject to right of Operator to abandon the well or to have the well completed at a lesser depth, Contractor agrees to drill the well to a total contract depth of_____feet. Contractor will drill the well on a footage contract basis (See Section 13a hereof) to_____feet, or to the top of the_____formation, or_____feet of penetration into_____formation, whichever is first reached. Drilling between the footage contract depth and total contract depth, if any, shall be at day work rates as specified in Section 13 hereof.

At Operator's request Contractor agrees to drill to a depth greater than total contract depth if in Contractor's opinion equipment at the well site is capable of such drilling. Rates for such drilling shall be negotiated by the parties hereto unless otherwise provided by Section 13 hereof.

4. RIG AND EQUIPMENT TO BE FURNISHED BY CONTRACTOR:

Contractor's Rig No._____
Drawworks:_____
Engines — Number, Make & Models:_____
Slush Pumps — Make, Model & Size:_____
Auxiliary Pump & Power:_____
Derrick or Mast — Make, Size & Capacity:_____
Substructure — Height & Capacity:_____
Drill Pipe — Sizes & Amounts:_____
Drill Collars — Sizes & Numbers:_____

Blow-out Preventers — Power Actuated:

Casing String	BOP Size	API Series	No. and Style	BOP Pressure Tests Frequency	Psi
Surface					
Intermediate					
Production					

Operational checks of BOP Equipment shall be made as follows:

(1)

FIG. 4–7 API bid sheet.

reputation of the drilling company. In planning, the operator must consider the dollar amount of the bids as well as the drilling contractor himself. What is his past performance record? Does he have the proper equipment, personel, and expertise? If the contractor making the lowest bid is not totally competent to do the job, then the lowest bid may not be the best.

Most operators use either the American Petroleum Institute (API) or the American Association of Drilling Contractors (AADC) standard form as the basis of their contract with the driller (Figs. 4–6, 7). As with the lease agreement with the owner of the land, the drilling contract sets forth the items of major concern to both parties. After both parties have been named, the contract identifies the location of the proposed well and establishes the date work is to commence.

Of major concern to both parties is the total planned depth of the well. If the driller is wise, he will not take a contract for a 10,000-foot well when he knows his equipment can only go to 3,000 feet. By the same token, a driller equipped for deep-well drilling is likely to waste time and money bidding on a shallow-well job. Accordingly, the operator should be familiar enough with the driller and his equipment to insure he is capable of completing the contract on schedule.

The same portion of the contract that covers the maximum depth also specifies the casing size and contains the "day-work basis" provisions. This sets out the amount to be paid by the operator to the driller. The first figure given is the cost per foot if the well is drilled on a footage basis. If the well is to be drilled on a day-work basis, the size of the crew is included and the daily (24-hour) and hourly rates are given for drilling, both with and without drillpipe. Also spelled out will be the standby time rate, or the amount paid to the contractor for remaining on the job even though no actual drilling may be in progress.

Other items specified in the contract will provide for the mud program, coring, sampling, logging, cementing, and testing if these services will be provided by the drilling contractor.

Components of the Rotary Rig System

Whether the modern rig is small, mounted on one or more trucks, or large, broken down and carried by a fleet, all rigs have three common essential equipment systems: the power equipment, the hoisting equipment, the rotary equipment, and the fluid circulation equipment (Fig. 4–8).

Power Equipment

Power requirements vary for different jobs depending on depth and rig design. However, most range from 1,000–3,000 horsepower (hp).

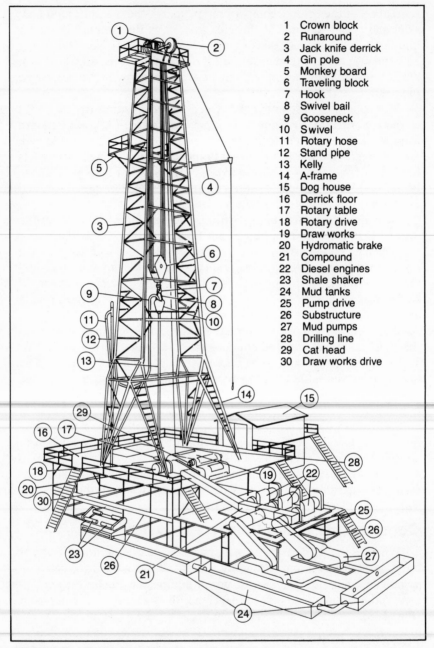

1	Crown block
2	Runaround
3	Jack knife derrick
4	Gin pole
5	Monkey board
6	Traveling block
7	Hook
8	Swivel bail
9	Gooseneck
10	Swivel
11	Rotary hose
12	Stand pipe
13	Kelly
14	A-frame
15	Dog house
16	Derrick floor
17	Rotary table
18	Rotary drive
19	Draw works
20	Hydromatic brake
21	Compound
22	Diesel engines
23	Shale shaker
24	Mud tanks
25	Pump drive
26	Substructure
27	Mud pumps
28	Drilling line
29	Cat head
30	Draw works drive

FIG. 4–8   Components of a drilling rig.

Shallow or modern rigs may be in the 500–1,000-hp class for hoisting and circulation, while rigs drilling 20,000 feet and deeper are usually 3,000 hp

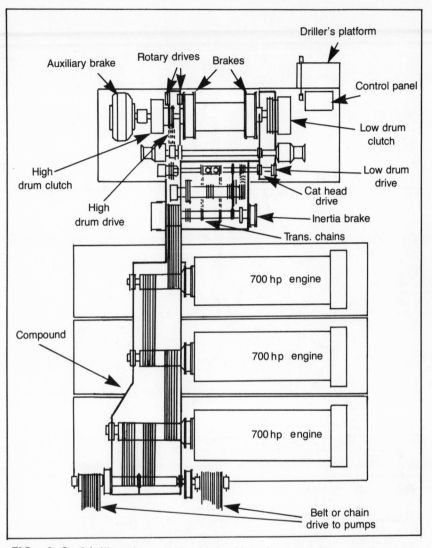

FIG. 4-9 Multiengine and chain-drive transmission used on a modern rig.

(Fig. 4–9). Auxiliary power for lighting, compressors, water supply, and additional needs is another 100–500 hp.

The main power source is usually one or more gas or diesel engines. Power is transmitted through chain drives to the draw works, rotary, and then through belt-driven drives to the pumps. Other rigs use diesel-electric or turboelectric power to generate electricity, which is routed through switching gear and cables to electric motors attached directly to the equipment.

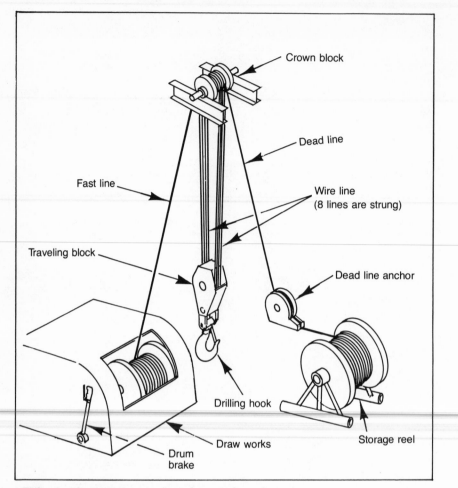

FIG. 4-10 Hoisting system for a rotary rig.

Hoisting Equipment

When drilling, the bit of the drillstring often becomes dulled and must be changed. When this happens, all of the lengths of pipe must be lifted out of the well, disconnected, and stacked upright outside the derrick's framework. This is called *tripping out*. The system that hauls up the thousands of feet of pipe and holds them suspended in the wellbore is called the hoisting system (Fig. 4-10).

One of the primary components of the system is the *derrick* or *mast*. These are rated in terms of how much vertical load they can carry and also how much horizontal wind force they can withstand. Derricks should be able to hold 250,000–1,500,000 pounds of pipe and can withstand a wind speed of 100–130 mph with the racks full of pipe.

The *draw works* is a revolving drum around which the drilling line is wound. Also on the draw works is the *cat shaft,* on which the *cat heads* are mounted. The draw works also houses the main brake, which must hold and stop the great weights of pipe being raised and lowered in the well.

There are two kinds of *cat heads* on the drum works. One is a drum; by using a large rope wrapped around it, crew members can use it to raise and lower loads from below the derrick onto the rig floor. The number of turns around the drum and the amount of tension applied to the rope creates the friction needed for the pull. The other cat head is a quick-release friction clutch and drum. The tong jerk line or spinning chain is attached to it and uses it for spinning up or breaking out the drillstring during trips and connections.

The *drilling lines* are wire rope ranging from 1⅛ inches to 1½ inches in diameter. One end, called the *fast line,* is attached to the drum of the draw works. The other end is connected to a storage reel located away from the rig area. When stringing up, or threading the cable through the rig, the *traveling block* is placed on the rig floor and the free end of the rope is threaded over the *crown block* and through the traveling block. The more times it ts wrapped around the two blocks, the heavier the load the derrick must support. The other end, called the *dead line,* is anchored.

The number of *sheaves* or grooves in the blocks through which the cable passes is always one more for the crown block than the number used on the traveling block. Thus, a 10-line string-up would use six sheaves in the crown block and five on the traveling block.

Rotating Equipment

The rotating equipment is all of the parts that begin at the swivel at the top and extend down the drillstring to the bit (Fig. 4–11). Each of the parts in this system of equipment moves and in turn provides motion for the bit.

The *swivel* carries the weight of the drillstring, permits it to rotate, and provides an entryway for the circulating fluid. Beneath the swivel is the *kelly.* This is a hollow six-sided tube that transmits torque to the drillstring from the rotary table and permits the string to move vertically as it is lowered during drilling. The bottom of the kelly fits into the *kelly bushing,* which is attached to the *master bushing*—the primary component in the *rotary table.* The kelly bushing allows the kelly to move up and down, yet fits into the rotary and transfers torque to the drillstring. The *kelly cock* is located above the kelly and can shut off back flow if a blowout occurs. The master bushing drives the kelly bushing and provides a place for the *slips,* tapered tools that hold the pipe suspended without having to use tongs.

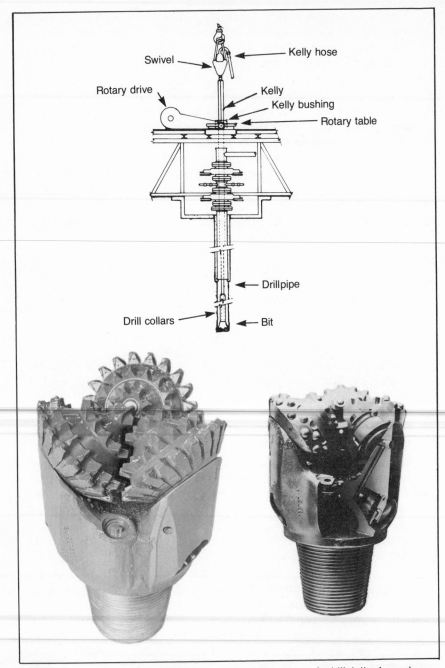

FIG. 4-11 The rotary system and close-ups of drill bits (courtesy Reed Bit Co.) A mill-tooth bit is on the left, and an insert bit is on the right.

The operating company in its contract with the drilling contractor carefully specifies the size and strength of the *drillpipe* used. Mud is pumped through the drillpipe (30-foot sections) and *drill collars* down to the bottom of the well. Drill collars are heavier than drillpipe and help lend stability during drilling operations. Rubber protectors for both pipe and collars are locked firmly into place to prevent metal-to-metal contact between the pipe and casing.

As with all the other items of drilling equipment, *bits* have undergone a steady evolution over the years. Basically, the job of the bit is to break up and dislodge the formation so that the drilling fluid can remove the cuttings. The conditions of the formation itself will determine the type of bit to be used. Three major kinds of bits are used: mill-tooth, insert, and diamond. Mill-tooth and insert bits are classified as roller-cone bits. Usually, 3 cones with teeth are attached to the end of the bit. These rotate, and the drilling fluid comes out in holes placed throughout the bit. The insert bit is much like the mill-tooth bit except that the teeth are made from a different material than the cone base. Diamond bits are for very hard formations. These are made by impressing diamonds into very hot metal and letting the metal cool. These bits, unlike the mill-tooth and insert bits, do not have cones or teeth.

Fluid Circulating Equipment

Fluid circulating equipment and mud represent one of the main costs involved in drilling a well.

At the beginning of the mud system are the *mud pumps* (Fig. 4–12). These machines pump mud into the *mud line* and pump water into the *mud mixing hopper*. Next, the mud goes to the *mud pits* where it is sucked up by the *suction line* and rises up the derrick through the *standpipe*. The standpipe is connected by the *mud hose* to the swivel. The mud in turn flows down through the kelly and the pipe, through the bit, and back up the annulus (the space between the casing and the pipe). Back at the top, it returns out the *mud return line* at the blowout-preventer stack and goes to the *shale shaker* to have the debris and cuttings removed. Then it settles out in the mud pit and is ready to be reused.

Drilling Fluid

The ancient Chinese added water to the borehole of their drilling wells as an aid to remove cuttings and soften rock formations. This practice was followed for hundreds of years of drilling throughout the world, especially for spring-pole and cable-tool methods.

Drilling fluid is a comparatively unimportant feature of cable-tool drilling because the cuttings are removed from the bottom of the hole with a bailer. After drilling a period of time the bit is removed from the borehole and a bailer is lowered to the bottom of the well to remove the

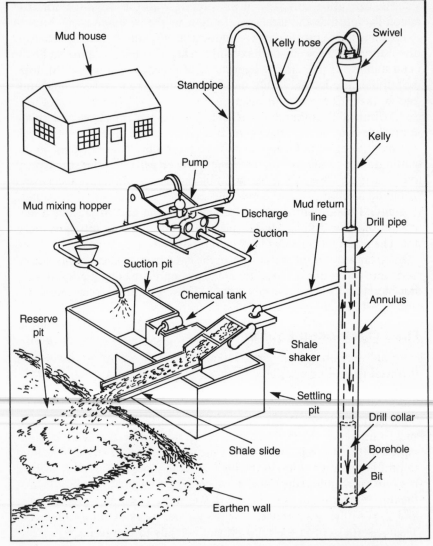

FIG. 4-12 Rotary-rig fluid circulation and mud-treating system.

cuttings. Water used as a drilling fluid causes cuttings to be in a slurry form and easily bailed out.

The early rotary drills used water as a drilling fluid. Under certain limited conditions water was effective since its function was only to force cuttings from the bottom of the well to the surface. It was soon realized that a better drilling fluid was needed because water could invade certain formations and block the pay zone. In 1889 John L. Buckingham tried to solve the problem. He mentioned the use of unctuous, or greasy, substances and oil-base muds in his patents as possible remedies for the action of water blocking.

Drilling mud development did not immediately follow the Spindletop discovery. Several years elapsed before the general realization that drilling mud control was important in the progress made on a rotary drilled well. Drillers relied on formation gumbo and sand to make drilling fluid (mud). If the mud got too thick, they added water. If the mud got too thin, they ran the risk of the drillpipe sticking, lost circulation, or a gas pocket blowing to the surface.

Some observant drillers did begin to foresee the need for methods and materials to control mud properties. They observed that sand was abrasive and damaged mud pumps. They discovered that a mud's density could be greatly increased by adding finely ground barite, hematite, or ground iron ore. As the density of the mud was increased, the hazard of a blowout was reduced and formations containing high pressures were drilled with a greater degree of safety.

It was discovered that the mud's viscosity rapidly increased when a clay called bentonite was added to the system. Bentonite is a predominantly montmorillonite clay that comes from Wyoming. Not all clays are desirable for drilling mud.

Clays are divided into four general mineral groups: kaolinite, illite, attapulgite, and montmorillonite. Neither kaolinite nor illite are desirable in drilling mud. Attapulgite clays have a specific use because of their property of forming a gel suspension equally well in salt or fresh water. They are used as viscosity builders in saline muds but alone do not satisfactorily reduce filleration. Montmorillonite clays are the most satisfactory for use in fresh-water drilling fluids. The majority of wells have been drilled with fresh-water muds.

Bentonite drilling mud that has been properly compounded will transport cuttings out of the hole and deposit a thin filter cake on the face of the formation that is relatively impermeable. It is responsive to chemical treatment for modifying viscosity and gel characteristics and suspends cuttings and weight material during circulation interruptions.

Mud Density

The density (weight) of drilling muds is usually measured with a mud balance or scale as shown in Fig. 4–13. These instruments are easily calibrated and are rugged enough to be used on-site in the field. Normally, one of the drilling crew will catch a sample of mud at various drilling depths and read its weight by use of the mud balance scale. By using the density of fresh water that has a known density of 8.33 pounds per gallon (ppg) (or 62.4 lb/ft^2) as a standard, it is possible to compare the sample mud density and get a reading. The general practice is to report mud density in pounds per gallon; hence a 12-pound mud is one having a density of 12 ppg. Hydrometers are also rarely used to calibrate mud weight, and calibration is based on the known density of fresh water.

Mud Viscosity

Viscosity is a measure of the internal resistance of a fluid to flow. The greater the internal resistance, the higher the viscosity. The viscosity of mud must be controlled and a standard means of measuring viscosity must be provided for proper drilling operations. Drilling rate, hole size, pumping rate, mud density, cutting size, and the design of the pit system are some of the factors influencing the specifications of the viscosity of any given mud.

Mud viscosity is difficult to measure. The development of satisfactory instruments for measuring the viscosity and gel strength of drilling muds has been the subject of much effort. The viscosity characteristics of some drilling muds cannot be described with a single measurement.

Routine on-site field measurements of the viscosity of drilling mud are made with a *marsh funnel,* as shown in Fig. 4–14. The operator collects a sample of well-agitated mud and holds the funnel in one hand with a finger over the tip of the tube. The sample mud is then poured through the screen end of the funnel until the level reaches the upper mark. The time is noted as the finger is removed and the mud is discharged into a one-quart (946-cc) container. The time in seconds that is recorded to fill the container is the apparent viscosity of the mud. Funnel viscosities are of little quantitative use but have general comparative value. Marsh funnels are manufactured to precise dimensional standards and may be calibrated with fresh water, which has a funnel viscosity of 26 ± 0.5 seconds with 1,500 cc in the funnel at a temperature between 70 and 80° F. The time is noted for one quart to drain from the funnel.

The direct-indicating viscometers as shown in Fig. 4–15 also measure apparent viscosity. They are multispeed instruments and better characterize the flow properties of mud. The apparent viscosity of a mud is composed of two variables, plastic viscosity and yield point.

Plastic viscosity depends on the concentration of solids present and the size and shape of these solid particles. It is that part of flow resistance in mud caused mainly by friction between the suspended particles and by the viscosity of the continuous liquid phase.

Yield point is a measurement under flowing conditions of the force in the mud that causes gel structure to develop when the mud is at rest. The gel strength of a fluid indicates the intensity of electrical forces existing between solid particles suspended in a fluid at rest. The particles tend to arrange themselves into a rigid, disorderly interlocked mesh or gel structure. The gel structure is destroyed with agitation yet will immediately

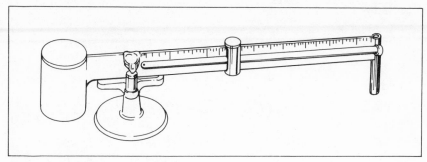

FIG. 4-13 Mud balance.

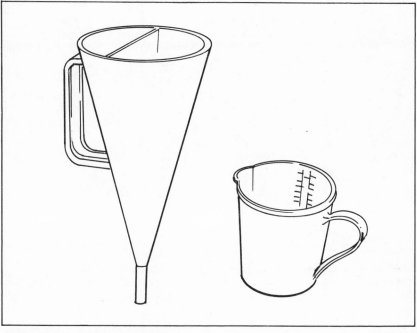

FIG. 4-14 Marsh funnel.

start to reform when the fluid comes to rest. This property of repeatedly becoming a fluid on agitation and again gelling when at rest is called *thixotropy*.

Basic Mud Functions

The general functions of drilling fluid are:

1. To remove and transport cuttings from the bottom of the hole to the surface.

2. To lubricate and cool the bit and drillstring.

3. To wall the borehole with an impermeable cake.

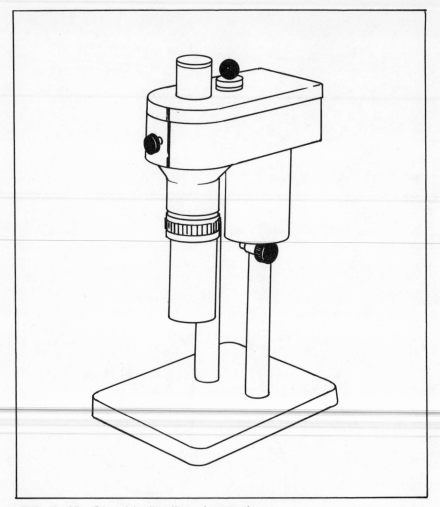

FIG. 4-15 Direct-indicating viscometer.

4. To control encountered subsurface pressures.

5. To suspend cuttings during times when fluid circulation is stopped.

6. To release sand and cuttings at the surface in the pits.

7. To support part of the weight of drillstring.

8. To reduce to a minimum any adverse effects upon the formation adjacent to the borehole.

9. To ensure maximum information about the formations penetrated.

10. To transmit hydraulic horsepower to the bit.

The general undesirable properties of drilling fluids are:

1. Could allow a high volume of filtrate to invade the formation.

2. Could tend to cause caving by swelling or hydrous disintegration of shales.

3. May reduce the oil permeability of a potentially productive formation.

4. May deposit a thick filter cake on the walls of the borehole that reduce hole diameter causing stuck drillpipe.

5. Could cause caving of the borehole if excessive deposit filter cake is present during removal of the drillstring.

Types of Drilling Mud

A broad classification of drilling mud is based on the composition of the liquid phase, such as fresh-water base, salt-water base, emulsion, and oil base. Muds may be further classified according to components, type of chemical treatment, or application. Drilling mud normally consists of the following fractions:

1. Liquid (water, oil, or both), which is the major fraction by volume.

2. Noncolloidal solids (sand, iron ore, barite, hematite), frequently the major component by weight.

3. Colloidal solids (clays, organic colloids), the major factor in determining the performance and properties of the mud.

4. Dissolved chemicals (mineral lignin, barium carbonate, sodium bicarbonate, formaldehyde, etc.) used to thicken muds and to control filtration.

Not all drilling muds mentioned above cover all of the drilling fluids used in rotary drilling. Air-gas fluids include compressed air, natural gas, mist, foam, and aerated muds.

ROUTINE PREPARATIONS

Few wells are drilled in ideal terrain. Therefore, the grounds must be prepared for drilling operations. First, a heavy equipment contractor prepares access roads. The well site itself is graded and leveled, and the surrounding area is prepared for equipment storage and crew quarters. Reserve pits must also be dug and diked, and a cellar is dug where the blowout preventer will be placed beneath the future rig floor. Wells may have to be dug or pipelines laid to bring in the water necessary for the drilling mud.

The framework of the rig is then installed and three holes are dug or driven with a pile driver before actual drilling operations begin. The first is the *conductor hole* where the conductor casing is set. This hole is dug from 20–100 feet deep, and it provides protection for the ground water aquifer. Another hole is the *rat hole*. This is also lined with pipe. During drilling, the kelly sits here when tripping in or out. The third hole, the *mouse hole,* holds the next joint of pipe to be used.

PERSONNEL

Many people are involved in drilling a well (Fig. 4–16). A large drilling contractor may have 40–50 rigs operating at the same time, scattered over the countryside. Typically, he establishes regional offices. The regional manager, the drilling superintendent, and the drilling engineer are usually responsible for about ten of the rigs.

Next in the chain of command is the tool pusher, who is in charge of the actual work site. Drilling is done 24 hours per day in three eight-hour shifts called *tours* (pronounced "towers"). The head of each tour is the driller, who is in charge of a derrickman and three roughnecks.

The *driller* is a special kind of person. He must be able to sense trouble before it actually occurs. In case of a blowout or accident, he must cope with the emergency and administer first aid. He must also keep up the morale of the crew and find a replacement in case one member becomes ill.

The *derrickman* perches himself high up in the derrick and is in charge of racking the pipe as it is pulled from the well. This can be a very precarious job due to high winds. Although he is connected to the earth with a cable called a *geronimo line,* there are times when he has no warning and is caught on the rig when an accident occurs.

The *roughnecks* or floormen are in charge of operations on the rig floor—one of the greasiest jobs in the business. They are responsible for screwing and unscrewing the joints of pipe as they come out of the hole and must wash down the pipe as it comes out. These crewmembers must enjoy working outside and must withstand the rigors of the climate from the Tropics to the Arctic.

CONTROL

From his position on the rig deck, the driller is able to see all of the necessary instruments and has within reach all of the controls for the various functions of the drilling process.

The most important consideration is the rate of penetration. This is where the driller's experience is called into play, for the proper rate of penetration for the particular formation is achieved by a combination of the weight placed on the bit, the torque (turning energy applied by the rotary), and fluid circulation.

The two most important controls are the power-plant throttles and

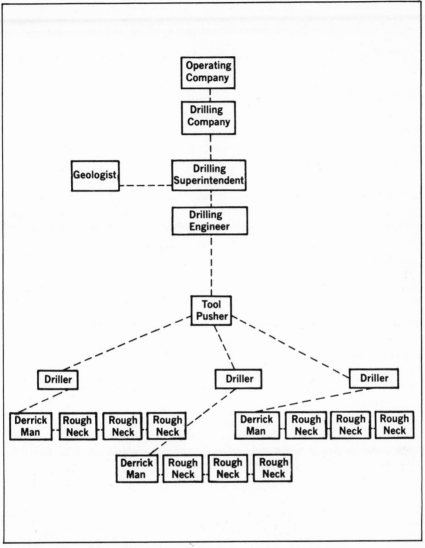

FIG. 4–16 Personnel organization for a drilling company.

the draw works brake, and the prime instrument for the driller to watch is the weight indicator. A pressure gauge enables the driller to determine the fluid pressure in the circulating system. It is usually placed near the weight indicator so he can monitor both the weight of the drillstring and the pressure within the circulating system at all times. Rotary speed is indicated by a tachometer on a gas or diesel rig or by an ammeter on diesel-electric rigs.

Fluid-flow sensors and indicating instruments allow the driller to monitor drilling fluid level losses or gains. Some of the instrumentation is equipped with both visible and audible alarm systems to alert the driller to unusual circumstances. By combining his knowledge and experience with the information shown on the dials and gauges, the driller is able to determine the best rate of penetration and can keep a close check on borehole conditions.

Also close at hand is a set of hydraulic controls for the blowout preventers so the driller can quickly close the preventers if abnormal conditions become apparent. A remote set is also usually provided away from the wellhead.

In addition to the instruments that provide direct read-out information, many operations are also monitored by recording instruments, which keep a permanent record on paper disks or tapes. Each keeps track of the date and time and a particular bit of information such as hook load, pump pressure, hole depth, rate of penetration, and fluid level.

ROUTINE DRILLING AHEAD

The next step in drilling the well is the actual drilling process. (Fig. 4–17). This is done one joint of drillpipe at a time. When making a connection (adding a new joint of pipe), the new pipe joint is raised from the storage rack and placed in the mouse hole on one side of the well. The swivel and kelly are disconnected from the string, swung over, and connected to the new joint, which is then raised into position over the bore and connected to the string in place.

Barring difficulties, drilling continues until it is time to change the bit, either because it has become dull or because a different type of bit is required. In order to change the bit, the driller must *make a trip*. Thus, a *round trip* would include bringing the drillstring out of the hole, replacing the bit, and getting the string back into the hole and into operation again. During the process, drilling fluid pressure must be maintained in the hole to prevent any possibility of a blowout.

The manner in which the trip is made will depend on the capabilities of the particular rig. Normally, the drill pipe is pulled out in strands of three joints, called a *thribble*. Smaller rigs can only handle a *double* (two joints), while a very large rig with a tall derrick can handle *fourbles* (four joints).

In *tripping out,* the kelly is drilled down before the drillstring is pulled. Then the swivel, kelly, kelly cock, and rotary bushing are placed in the rat hole out of the way. The elevators on the hook and block assembly are latched around the pipe just below the tool joint box and the pipe is then pulled and stood on the rig floor until time to return the string (Fig. 4–18).

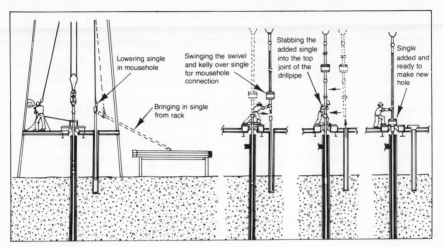

FIG. 4–17 Adding a new joint of pipe.

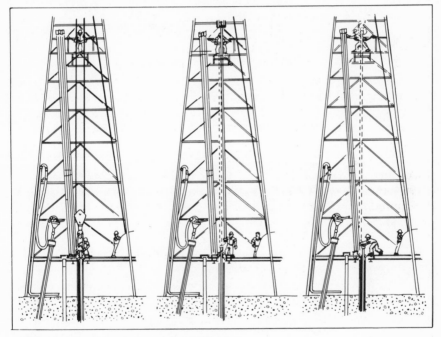

FIG. 4–18 Tripping out the drillstring.

CEMENTING AND CASING STRINGS

As the well is dug deeper and deeper, it finally reaches below the depth of the conductor hole. At this point, conductor casing must be set to protect the fresh-water formations (Fig. 4–19). Large-diameter pieces of casing are lowered and cemented into place to keep the walls of the formation back. When the casing is in place, a second string of casing is placed down inside it called *intermediate casing*. This runs from the surface down to the production zone. When the intermediate casing reaches the bottom of the surface casing, cement slurry is pumped down the intermediate string of casing after a bottom plug and is guided back up the annulus between the intermediate and surface casings with help of the guide shoe (Fig. 4–20). When the cement has completed its journey, a top plug is introduced at the surface. It pushes out any remaining cement in the intermediate string and forces it back up the annulus. The result is a cement bond between the two casings and an open well for continued drilling operations.

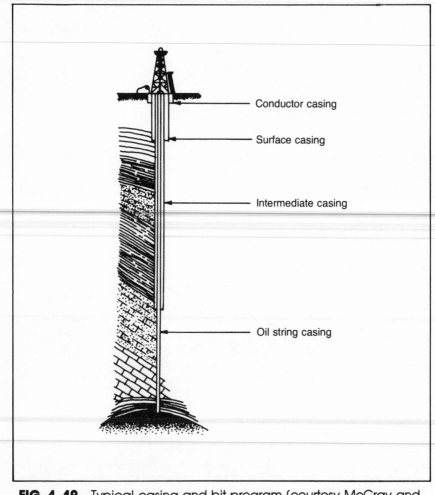

Conductor casing

Surface casing

Intermediate casing

Oil string casing

FIG. 4–19 Typical casing and bit program (courtesy McCray and Cole, University of Oklahoma Press, 1959).

SPECIAL DRILLING

As drilling continues, other kinds of tools are used to keep the casing in the middle of the hole. *Centralizers* and *scratchers* keep the casing in the middle of the hole and prevent the casing from *keyseating* (Fig. 4–21). This is when the casing becomes stuck on one side of the hole and must be reamed free. If this happens, a device called a *free-point indicator* is lowered into the hole. It locates the stuck joint, and then a *string shot* is lowered to that point. This instrument disconnects the drillpipe above and allows it to be raised from the hole. Over the stuck pipe are

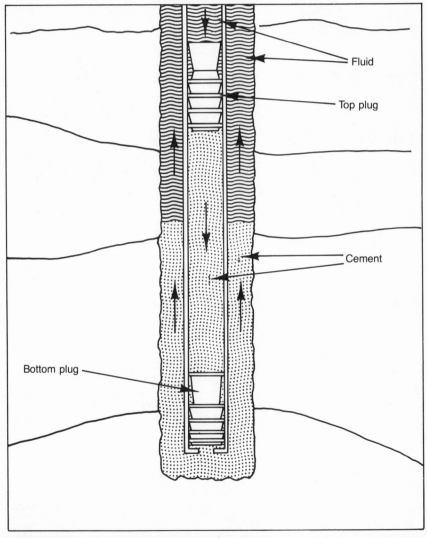

FIG. 4–20 Casing cementing procedure.

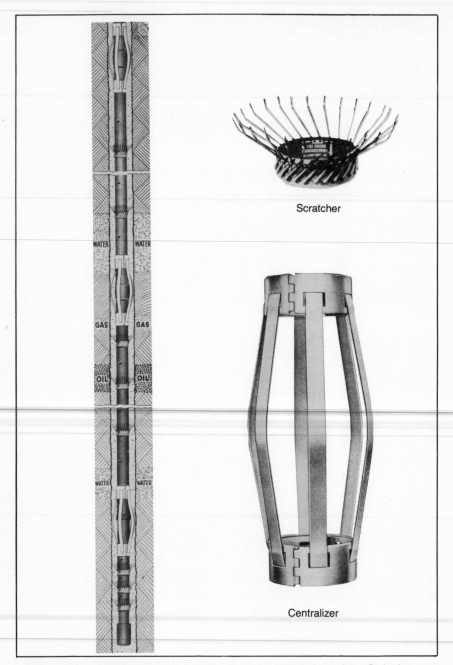

Scratcher

Centralizer

FIG. 4–21 Scratchers and centralizers in place (courtesy Petroleum Extension Service).

lowered a *washover pipe,* which closes around the dangling string and grinds up the wall cake (the material that forces the pipe to stick), and a *back-off connector,* which screws into the stuck tool joint and keeps it from falling to the bottom of the well once it is freed.

Other times, a *bumper jar* can be lowered into the hole to rescue stuck pipe. A spring inside the tool is contracted, and the tool is lowered to the stuck pipe and connected. Then the spring is released. Depending on the direction the driller wants the pipe to go, the jar can either move the pipe up or down.

Sometimes, unfortunately, either pieces of pipe or tools are lost down the hole. These are called *fish,* and tools have been devised to

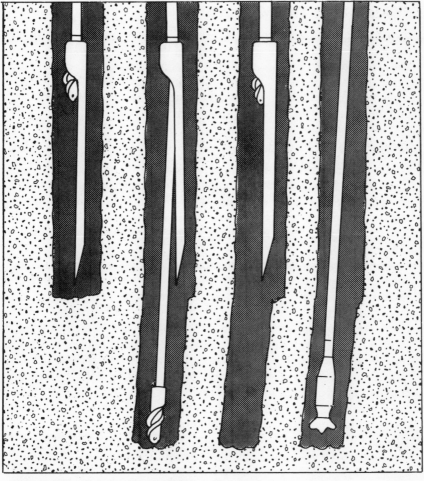

FIG. 4–22 Sidetracking with a whipstock.

locate the equipment and pull it out of the well. However, this is costly and time-consuming. So sometimes the driller decides to *sidetrack*. He lowers a tool called a *whipstock* downhole. This angles the pipe in a new direction (Fig. 4–22).

Occasionally, a driller will want to change his direction because tests show he is off course. There is always some deviation from the vertical in every well. The weight of the drill collars is usually enough to keep the hole on course. Sometimes, though, a special electrical instrument is lowered into the hole. It has a meter that reads the direction and automatically rotates the drill bit in the right direction. This method is very expensive and is only used in formations that will yield good pays.

EVALUATING THE FORMATION

At this point, the operator decides whether or not the well will be profitable enough to set production casing. By testing cuttings and cores and with the help of logs and other wireline tests (Fig. 4–23), he can determine whether or not a profitable zone exists downhole. If it does, he will install production casing and cement it in place. If not, he will plug the well, and it will be written off as a *dry hole*.

COMPLETION

Depending on whether or not the well will be profitable, the contractor may now either run tubing, set the wellhead, and proceed to bring in the well, or after the casing is set and cemented he may move off the site and a well-service operator will bring his rig in to complete the final operations. As with the other steps, the deciding factor is the cost to the operator.

Economics also dictate the method of completion selected. The most widely used method is perforated casing. This means puncturing through the tubing, cement, and casing into the formation so the oil or gas can enter the wellbore. If a perforating gun is used, a tool much like a wireline instrument is lowered to the bottom of the hole. Steel projectiles (bullets) are fired through the casing into the formation, forming tunnels for the liquid to migrate into the well (Fig. 4–24). The second method, shaped-charge, uses small gas explosions to drive a hole through the cement and casing and into the formation.

If only one pay zone is encountered, the well is called a *single completion* (Fig. 4–25). If two different zones are encountered, two tubing strings are lowered and are separated with a rubber packer that keeps the fluids from the two different formations from mingling (*multiple completion*).

Some formations are very sandy, and the granules keep clogging the well. When this happens, charcoal filters or screens are used, which keep the well clean and the oil flowing freely. In other formations, the rock is

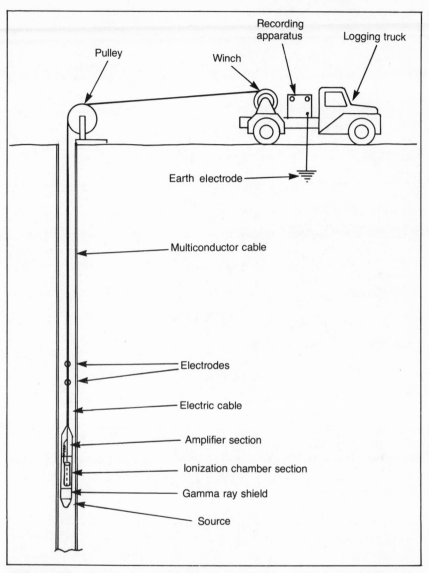

FIG. 4–23 Placing a wireline instrument downhole for well testing

so solid that no cementing is necessary. This is called an *open-hole* or *barefoot* completion. In this process, the fluids flow naturally into the wellbore and up into the tubing.

Acidizing and Hydraulic Fracturing

At some point, it may be necessary to open up the formation to increase the flow. If the formation is composed of carbonate rocks and

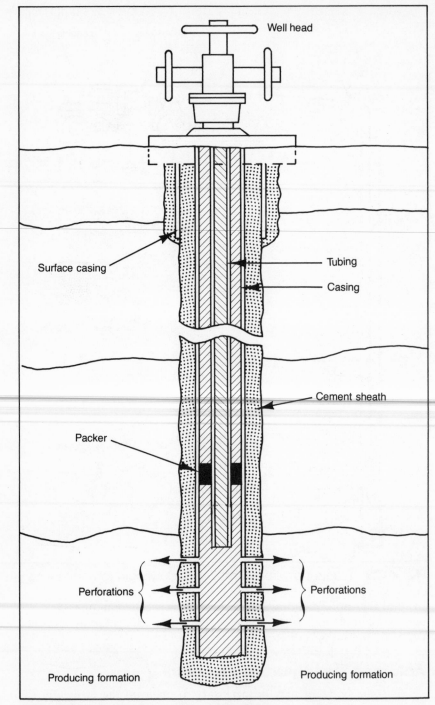

Well head

Surface casing

Tubing

Casing

Cement sheath

Packer

Perforations

Perforations

Producing formation

Producing formation

FIG. 4–24 Perforated casing completion.

the permeability is low, then acid will open paths in the rock for the oil to flow through, creating greater porosity (Fig. 4–26). This again calls for the services of a special contractor who will pump acid under great pressure down the well and out into the formation.

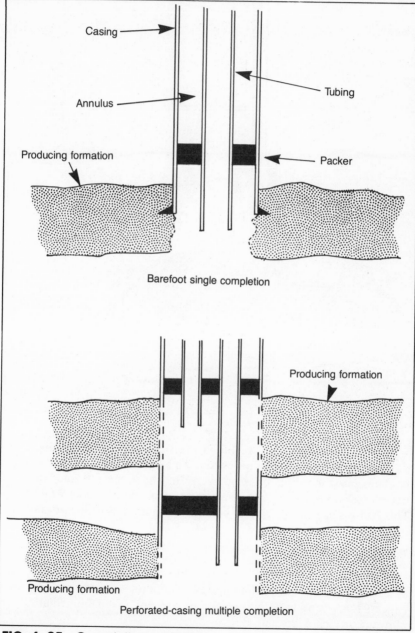

Casing

Tubing

Annulus

Producing formation

Packer

Barefoot single completion

Producing formation

Producing formation

Perforated-casing multiple completion

FIG. 4–25 Completion methods.

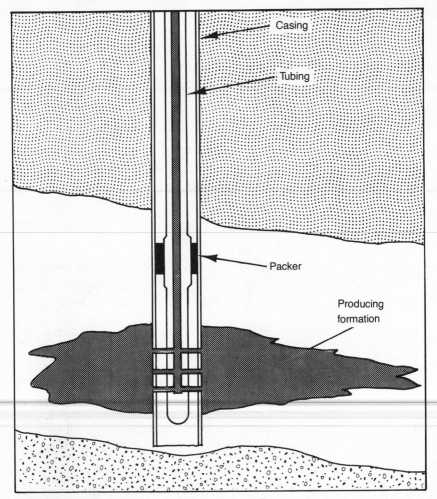

FIG. 4-26 Acidizing and fracing a well.

If there is oil and gas available in sandstones but the permeability is too low, a *frac* job may be necessary. This involves forcing a sand and fluid mix into the formation to open cracks. The particles of sand remain in the cracks, keeping them open so the hydrocarbons can flow through. This is also a specialized service calling for technicians with custom-built pumps and equipment.

WELL CONTROL

The area of well control deals with blowouts. At all times, the pressure downhole is closely monitored. Sometimes, however, the unexpected happens and, if precautions are not taken immediately, a blowout or kick will occur (Fig. 4-27).

FIG. 4–27 A blowout (Billie Linduff collection).

Blowouts usually have definite signs: shows of oil and gas in the drilling fluid, a sudden increase in the drilling rate, increased flow of mud into the mud pits, kicks of drilling mud flowing from the well casing, and lightened drillstem weight.

Kicks and blowouts all stem from imbalances in formation pressure. Formation pressure is the force exerted by fluids in a formation. It increases with depth, i.e., the deeper the well, the greater the pressure. This is why fluids flow from a well—fluids flow from the area of greatest pressure to the area of weakest pressure.

FIG. 4-28 A blowout preventer stack, also known as a BOP stack.

The drilling mud that cools the bit and removes the cuttings is the best way for the driller to prevent the formation fluids from rising to the surface prematurely. Sometimes, though, very porous or fractured formations are encountered. When this happens, some of the drilling fluid seeps out into the formation (a *zone of lost circulation*) and the pressure drops, paving the way for a blowout. At other times, an area of abnormally high pressure may be encountered and the weight of the drilling mud will not be enough to hold it back. For another case, the swabbing or sucking action of the drillstem as it is being pulled from the hole during bit changes draws drilling fluids up the well. This decrease in pressure downhole is just enough to give the formation fluids the space they need to expand upward.

To help prevent these kinds of problems, *blowout preventers* (BOPs) are installed at the top of the wellhead (Fig. 4-28). These close the top of the hole in case of a blowout, control the release of fluids, permit drilling mud to be pumped down into the well, and allow the drillpipe to be moved. If a blowout occurs, three sets of rams can be used to close in the well. These seal off the annular spaces and can often prevent the fluids from venting to the surface. The best preventative, though, is careful monitoring.

CHAPTER 5

Offshore Drilling

The objective and many of the methods of drilling an offshore well are the same as for one drilled on land. But the complexities, equipment, expense, and personnel are greatly increased.

One of the earliest patents for an "offshore" drilling rig was issued to T. F. Rowland in 1869. While it was for use in shallow water, its anchored, four-legged tower was the forerunner of today's platforms. The first actual offshore operations in the U.S. began in 1886 when a coastal field was discovered in Santa Barbara county, California. The first well was drilled from the shore, but by 1890 wells were being drilled into the Pacific from wharves built out from shore, some as far as 1,200 feet from the beach. More than 200 such wells were drilled, but production was low and the rigs, placed on pilings, were susceptible to storm damage.

Then close-to-shore wells were drilled in the Gulf of Mexico along Texas and Louisiana. In the 1920s, wells again began to be drilled from wharves and piers in California. In 1932, the first platform-supported well was drilled at Rincon. At the same time, offshore wells were being drilled directionally from shore locations at Wilmington and Huntington Beach. By the end of that decade, wells were being drilled off the Louisiana coast in the Gulf.

During World War II, the military borrowed offshore expertise and provided new knowledge of their own in offshore construction when the famous Texas Towers were built. These platforms, much like those used for drilling operations, were built to house radar units to warn of impending enemy attack along the coasts of both the United States and Great Britain. Some of those off England, abandoned by the authorities after hostilities ceased, became the homes of "pirate" radio stations that competed with the government-owned broadcasting stations in Europe.

After World War II, it was realized that offshore areas would provide rich new fields, and since then thousands of wells have been successfully drilled despite the special problems encountered. Land-based operations and methods have been adapted where possible and new marine techniques developed in exploration, drilling, production and transportation.

EXPLORATION

Much petroleum exploration has been done underwater on the continental shelves that lie off the world's coasts. Basically, the situation is similar to that on land. The potential areas are in sediments laid down millions of years ago. Seismology is used in the search for traps (Fig. 5–1). A ship tows a sound source through the water. Echoes from the sound pulses sent out every ten seconds are picked up by shock detectors spaced along a cable towed behind the ship, just as geophones pick up the echoes on land. Depth charges were first used to produce the necessary shock waves, but these have been replaced with air bursts that are not harmful to marine life.

DRILLING SITES

The biggest difference between onshore and offshore operations is that the actual base of drilling operations is man-made. When exploratory wells are drilled at sea, the procedure is more complicated than it is

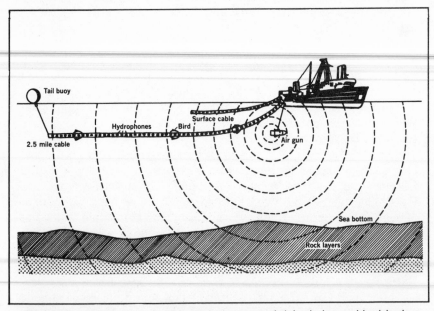

FIG. 5–1 Seismic operations at sea use air blasts towed behind a ship to create shock waves that are picked up by hydrophones and transferred to the ship (courtesy Mobile World).

on land where the surface provides a base for the drilling rig. Therefore, depending on the depth of the water, the climatic conditions of the area, and the costs involved, an operator can choose from several different types of drilling structures.

Drilling Barges

For inland and shallow-water areas, a drilling rig mounted on a barge may be used (Fig. 5–2). Barges differ from ships, in that they are not self-propelled but must be towed by a tug. They are designed to be used in shallow water in a bay or estuary. However, the wave action in deep water would submerge them. In addition to the necessary drilling equipment, barges also house supplies and crew quarters. Completion services, such as cementing, may be furnished from other ships or barges especially equipped for that purpose.

FIG. 5–2 Typical drilling barge (courtesy Noble Drilling Co.)

Submersible Rigs

Submersible rigs are among the oldest of mobile exploratory drilling rigs (Fig. 5–3). Their obvious advantage is that they can be moved when the exploratory work is completed. The unit consists of two hulls. The upper hull contains the crew's quarters and the working area. The lower hull provides the buoyancy when being towed to the site and is then flooded to provide a stable base on the ocean floor. However, they are limited to relatively shallow water.

Jack-up Rig

This kind of rig can also be towed to the drilling location (Fig. 5–4). At the site, the legs are lowered to the sea bed and the platform is jacked

FIG. 5–3 Typical submersible rig (courtesy Noble Drilling Co.)

up to a safe level above the sea. Since it, like the submersible, rests on the sea floor, the jack-up is limited to relatively shallow depths. Unlike the submersible, though, it provides more area between the ocean and the working surface, making operations safer.

Semisubmersible

These vessels are the most widely used type of rigs (Fig. 5–5). Some of these can sit on the bottom in shallow water, but they are more fre-

FIG. 5–4 A jack-up drilling rig.

quently used in a partially submerged position, moored by massive anchors weighing nearly 22 tons apiece. Once the well has been completed, the semi can be refloated and moved to the next drilling site.

Drillship

In very deep waters, the drillship is used (Fig. 5–6). This is a ship with an opening through the hull for drilling. It is usually moored in the same manner as a semisubmersible. However, drillships supplement or replace the mooring system with computer-controlled motors mounted

FIG. 5–5 A semisubmersible drilling rig.

around the hull. These motors, called *thrusters,* compensate for the motion of the waves. The entire technique is called *dynamic positioning.* Like the semi, when the drilling is completed, the drillship moves off to another location.

FIG. 5–6 The Douglas Carver, a drillship.

Permanent Platforms and Drilling Islands

In some environments where the elements are particularly cruel, an entire island may be constructed for drilling (Fig. 5–7). Sand and gravel are carried by ship from the land and deposited on the ocean floor. Eventually, a man-made island is formed where stable drilling operations can be performed.

FIG. 5–7 A drilling island used for production and drilling in very severe conditions close to shore.

PROBLEMS OF OFFSHORE DRILLING

From exploration to production, the recovery of offshore hydrocarbons presents unique problems not encountered in shore operations. First, crew schedules are entirely different from onshore. Offshore, tours are normally 12 hours on and 12 hours off with a seven-day break every seven days. Offshore crews are also much larger and sometimes are far from land. Therefore, living quarters like large hotels, laundry service, meals, and entertainment must be provided for the men. If a drilling ship or barge is used, a ship's crew is also needed.

In addition to the catering and laundry services, all other supplies must also be brought from shore. Large items may be delivered by work

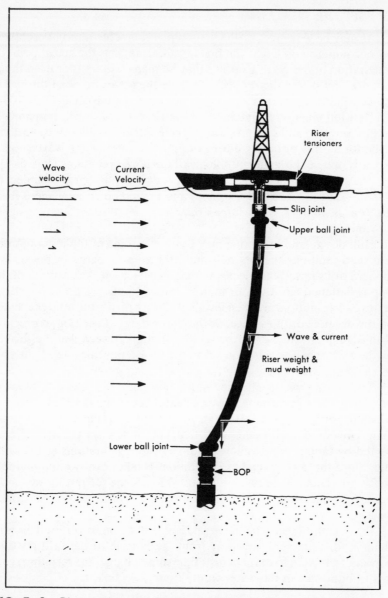

FIG. 5-8 Riser system for floating drilling rig (from OTC 1018).

boats with crew boats handling personnel transfers. And helicopters have become essential for fast transportation.

Another problem of offshore operations is maintaining uninterrupted communications between shore and rig. Telephone service is unavailable, so radio telephone is used instead on special frequencies assigned for that purpose.

OFFSHORE DRILLING

As may be expected, drilling at sea presents some unusual problems not encountered on land. The first is, of course, that the actual point of penetration of the earth's surface may be many hundreds or even thousands of feet below the rig deck. Drilling then must be done through a device called the marine riser, which extends from the ocean floor to the rig. Tension must be maintained on the riser so that it will remain in a vertical position and stay properly aligned. However, allowance must be made for motion of the rig during drilling due to wind and wave action, which is one of the major problems of offshore operations (Fig. 5–8).

The surface of the sea is never still. A drill ship or barge is subject to side-to-side roll, fore-and-aft pitch, and up-and-down movement from the sea as well as wind forces that may be constantly changing in direction.

Hull design and ballasting may overcome some of the motion problem, and multiple anchors help hold the ship or barge in place. For modern drillships, dynamic positioning may be used. The bottom of the ship is flattened and thruster motors and hydrophones are installed. A series of transmitters are then installed along with the blowout preventer on the ocean bed. In essence, as the ship moves off position, correction data is automatically sent from the transmitters, received at the surface by the hydrophones, and relayed to the thrusters, which apply the correct amount of compensating force.

The marine riser is attached at the bottom to the undersea blowout preventer. This in turn may be attached to a drilling template, through which the drillstring will pass to its proper position (Fig. 5–9). The template consists of two plates with matching holes, separated and connected by tubing, which allow multiple drilling operations at the same site. Since the cost of offshore operations is fantastically high, multiple wells are undertaken wherever feasible. Some offshore sites may become production centers for as many as 70 directionally drilled wells.

Provisions may also be made on the sea bed for a diverter. This is to channel high-pressure gas away from the bottom of the drill rig. If a pocket of gas were to erupt beneath the ship or barge, the resulting rapid loss of buoyancy in the water would cause it to sink.

SAFETY

Safety requirements at sea are even more stringent than those on land. Barges and ships must meet Coast Guard regulations pertaining to fire protection, life saving, and navigation. Fire lines, fire extinguishers, life-saving equipment and other safety apparatus are subject to periodic

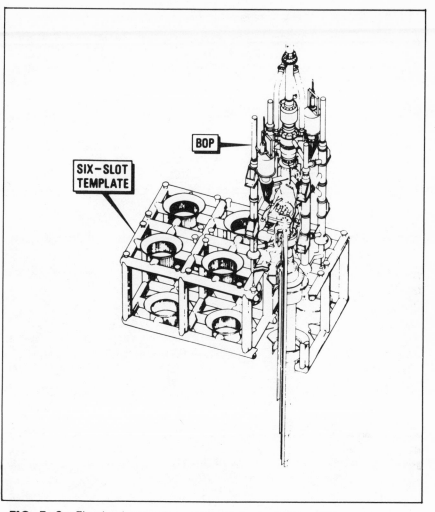

FIG. 5-9 Floater-type subsea template (courtesy JPT).

Coast Guard inspection. Life vests, protective clothing, and emergency signaling equipment must be furnished and maintained.

Watertight escape capsules are provided for use in the event the rig must be abandoned. Despite all the precautions and safety equipment, high winds, slippery steel decks, and the fury of the sea itself create hazardous conditions, and a number of lives are lost each year.

On deep-sea rigs, it may also be necessary to provide decompression chambers for the divers aboard. Some are now working at such great depths it may take them several days to decompress, and they virtually live in the chambers between dives.

ENVIRONMENTAL CONCERNS

In addition to all the other natural and legal constraints, offshore operators are faced with the problem of producing much-needed energy without harming the environment. Concern for our environment in recent years has focused much attention on the oil industry. Today most drilling contractors and operators are extremely scrupulous in their attempts to conduct clean operations (Fig. 5–10). Blowouts and oil spills

FIG. 5–10 An oil-spill boom helps contain oil to a specified area so it can be cleaned more easily.

are very costly in terms of lost production, damaged or destroyed rigs, and clean-up and salvage operations, as well as the resulting bad press the operator might receive. Despite their caution, however, there is the possibility that human error or a natural disaster could bring about a spill of major proportions. Every step is usually taken, though, to preclude such an unwanted event.

Drilling muds are occasionally discharged into the sea, but the total of such materials so dumped is far less than the amounts of solids carried into the oceans by the world's rivers each day. This process may actually enhance marine life. The muds tend to create artificial reefs on the ocean floor, and barnacles and other organisms attach themselves both to the residue and to the legs of the rig itself. They attract predators, who in turn attract larger fish. In fact, many sports fisherman claim some of their best catches are made near rigs.

CHAPTER 6

Production

Very few of the wells completed flow free of their own accord. Those that do are topped with a Christmas tree (an array of valves and chokes) that connects the well via a pipeline to the tank battery. Approximately 90 percent require some means of artificial lifting system to bring the wanted hydrocarbons to the surface. The method chosen depends on the depth of the well, the nature of the sands, the gas-oil ratio, the viscosity of the crude oil, and of course the cost involved. Lifting methods and machinery fall into one of two general categories: surface and subsurface.

SURFACE LIFTS

The first gusher in the United States was, of course, Spindletop. The wells drilled in the Drake era, since they did not flow to the surface, required some means of bringing the oil from its natural level in the wellbore on up to the top of the wellhead. Thus lifting methods developed concurrently with drilling methods ever since that point in history.

Drake's well was not a gusher and the oil did not flow upward out of the bore; therefore, it had to be lifted out, as did the oil from the others of the time. Drake and Uncle Billy Smith devised a bailer system that was both crude and slow but which did suffice for their production of 30 barrels a day. It was almost immediately apparent, however, that a better way was necessary.

Rod Pumping

By 1869, the rod pump system was in wide use in the early fields. Actually, all this involved was leaving the standard cable-tool drilling rig on the well site and letting the walking beam run the pump. In the past, rod-activated pumps had been used on the salt brine wells, so the idea was not really new. Wooden sucker rods with wrought-iron fittings were used at first. The walking beam gave the reciprocating movement, which moved the rod up and down, activating the pump. This principle, almost

2,000 years old, has evolved into what has become known as the standard pumping rig. Like other oil-field equipment, it has been improved over the years.

Today, the "horse head" bobbing up and down is a familiar sight in oil fields around the world and this method of bringing oil to the surface (or a variation of it) accounts for some 80 percent of the artificial lifting done (Fig. 6–1). Originally powered by steam, the pump may also use an

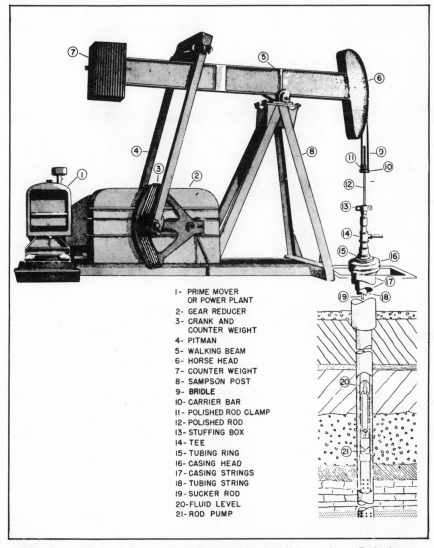

1- PRIME MOVER OR POWER PLANT
2- GEAR REDUCER
3- CRANK AND COUNTER WEIGHT
4- PITMAN
5- WALKING BEAM
6- HORSE HEAD
7- COUNTER WEIGHT
8- SAMPSON POST
9- BRIDLE
10- CARRIER BAR
11- POLISHED ROD CLAMP
12- POLISHED ROD
13- STUFFING BOX
14- TEE
15- TUBING RING
16- CASING HEAD
17- CASING STRINGS
18- TUBING STRING
19- SUCKER ROD
20- FLUID LEVEL
21- ROD PUMP

FIG. 6–1 Conventional beam pumping unit (courtesy Petroleum Extension Service).

internal-combustion engine or an electric motor. Those that are electrically powered include a reduction box or countershaft in the power train to slow the motor's rpm down to that needed to operate the unit.

Variations of the standard rig have included the air-balance beam pumping unit (Fig. 6–2.) This uses air pressure against a piston in a cylinder to counterbalance the weight of the fluid being lifted instead of a weight attached to the beam. This system allows for changing the counterbalance by merely adjusting the air pressure in the cylinder and also eliminates much of the mass of the unit and its foundation.

FIG. 6–2 Air-balanced beam pumping unit. The compressed air at point A counterbalances the weight of the sucker rods. Air pressure is furnished by a compressor at B.

Another variation is the cable-operated long stroke (Fig. 6–3). This unit, instead of having the standard walking beam, is mounted on a 50-foot tower erected over the well and has a 34-foot stroke, much longer than any standard rig. Since there are fewer cycles, the unit is said to lift more oil with fewer rod failures. Because the stroke is longer and speed more uniform, a more viscous fluid can be lifted and the downhole efficiency is increased.

Still another type is the Lufkin Mark II. Like the air-balance beam unit, the crank is mounted on the front and utilizes an upward thrust instead of a downward motion to raise the sucker rod. The gear box and

FIG. 6–3 Cable-operated, long-stroke vertical lift pump (courtesy Petroleum Extension Service).

crank are so arranged that the beam has a slow upward stroke and a fast downstroke. This, the manufacturer says, increases production per stroke, lowers operating costs, and makes it possible to use a smaller power unit.

Introducing a new generation of energy-saving pumping units is the Alpha I. Essentially a winch-and-cable system, it operates like a lightweight draw works to raise and lower the rod string in the well. The Alpha I uses only three strokes per minute as compared to 12 to 15 strokes on a well of similar depth. Another advantage cited is that the long, slow stroke of the unit extends rod string life and improves the efficiency of the bottom-hole pump, which results in reduced maintenance. Energy is saved because the electric motor turns off on the downstroke. When the motor stops, the unit rolls to a stop and a cam starts the unit rolling in the opposite direction, then the motor restarts. Thus the motor only runs five-eighths of the time instead of full time.

The heart of the standard pumping unit is the sucker rod, which does the actual job of lifting the oil to the surface (Fig. 6–4). Even though it is a rod, it acts much like a flexible spring and operates under great stress. The sucker rods can be easily damaged by improper handling, and any bends, nicks, or dents can lead to metal fatigue and early failure. If the pump sticks, the rod may be stretched beyond its elastic limits and break. Therefore, great care should be exercised when running or pulling rods, and they should be kept stacked in a rack for that purpose so no bending will occur.

At the bottom of the sucker-rod string is a piston or plunger pump submerged in the fluid of the well. As the sucker rods are drawn up by the beam unit at the surface, the pump opens and allows the fluid to enter the pump chamber. As the rod closes, the fluid inside is forced up through a valve into the tubing, with another valve closing behind it to prevent escape back into the formation. Then the process is repeated. Slowly but steadily, the fluid is raised to the surface.

SUBSURFACE LIFTS

Another method of bringing wanted hydrocarbons to the surface is by means of an electrically powered submersible pump (Fig. 6–5). A centrifugal pump, together with its driving motor, is lowered down the bore to the bottom of the well. Power is transmitted through an electrical cable from a surface control box. The pressure created by the rotation of the pump's impellers then forces the fluid to the surface.

When first introduced, problems arose from the failure of the insulated power cable. But as with all oil-field equipment, efficiency and reliability has constantly been improved. A typical submersible pump may lift from 250 to 25,000 barrels per day, depending on the size of the casing and the depth of the well.

Subsurface hydraulic pumping actually uses the oil in the field to force additional oil to the surface (Fig. 6–6). This too uses a pump at the bottom of the well driven by a hydraulic motor. On the surface, a standard engine-driven pump draws clean crude oil from the top of a settling tank and forces it down through tubing to the hydraulic motor. The oil,

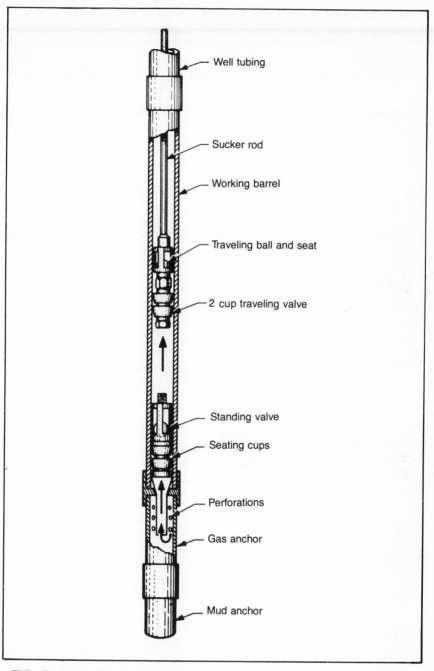

FIG. 6–4 Sucker-rod pump.

under pressure, drives the motor, which in turn drives the pump. The driving oil is exhausted from the motor into the well where it mixes with

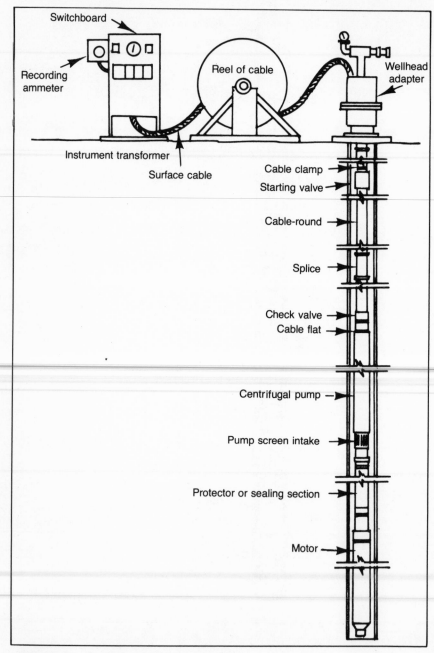

FIG. 6–5 Centrifugal pumping unit.

the oil to be pumped and the pressure from the pump lifts both to the surface through a second string of tubing. One barrel of power oil will usually lift one barrel of "new" oil (oil from a recent well).

In fields where a large supply of gas is available and the amount of fluids expected to be recovered justifies the cost, gas lift may be used (Fig. 6–7). Gas is introduced into the well in the space between the tubing and the casing. A series of gas-lift valves are in place along the

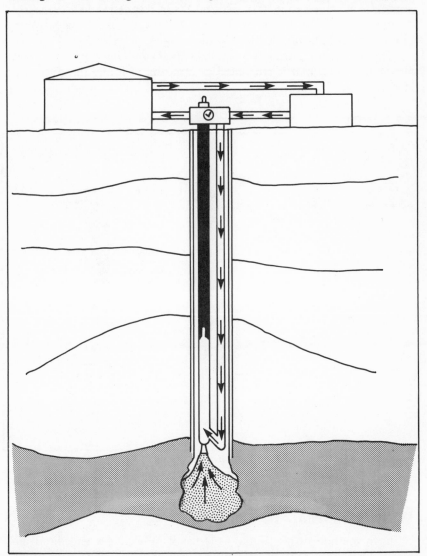

FIG. 6–6 On a hydraulic pumping system, power oil is pumped downhole and turns a motor at the bottom of the well.

tubing, and these close as the gas enters the next lowest valve. As the gas enters the crude stream, it aerates the thick oil and lightens it to move more quickly to the surface.

Other methods of lifting apparatus that have been tried include the sonic pump and the ball pump. As with the gas lift, a series of valves are installed in the tubing string which is suspended from a wave generator. Oscillations are produced that proceed down the tubing at the speed of sound (as it travels in metal). This results in resonation in areas of the tubing, causing it to expand and contract. This makes the check valves open and close, lifting the fluid to the surface.

The ball pump utilizes two parallel strings of tubing through which a synthetic rubber ball is circulated as a means of lifting the oil to the surface.

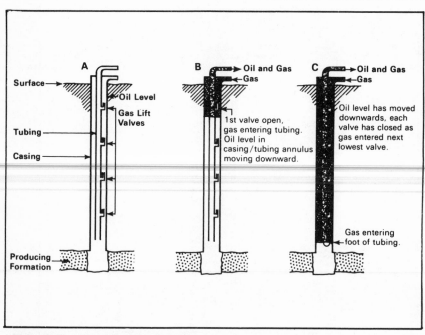

FIG. 6–7 Gas-lift artificial lift.

WHICH METHOD TO CHOOSE?

Each method of lift (Table 6–1) has its advantages and disadvantages. Therefore, the operator of the well or field will have to weigh these carefully before selecting the method that would work best for his particular situation.

TABLE 6-1 Choosing a Method of Lift

Method	Advantages	Disadvantages
Free Flow	Produces by natural reservoir drive. No artificial lift necessary.	
Standard Beam Rig	Proven and improved over many years of use. Motive power available from gasoline, diesel fuel, natural gas, or electricity. May be the cheapest lift and its operation is familiar to field personnel. Pumps can usually handle sand or trash. It is possible to pump off.	Sucker rods subject to failure. Unit required at each wellhead on lease. Volume decreases as depth increases. Rods must be pulled to change pump. Unit is susceptible to free gas in pump.
Hydraulic Pumping	High volume can be produced from great depth. Power equipment on the surface can be centralized. Pumps can be changed without pulling tubing. It is almost possible to pump off.	Susceptible to free gas in pump and is also vulnerable to solids in pumps. Well testing is difficult because power oil is also in well stream. This also causes oil treating problems.
Submersible Electric Pumping	Very high volumes at shallow depth can be produced. It is almost possible to pump off.	Maximum volume drops rapidly as depth increases. Very susceptible to free gas in pump. Tubing must be pulled to change pump and cable. Unit required for each well.
Gas Lift	Takes advantage of the gas energy in the reservoir. Is a high-volume method. Can easily handle sand and trash. Valves may be retrieved by wireline or hydraulically.	A source of gas must be available. It cannot pump off. Minimum bottom hole pressure increases with depth and volume.

WELL TESTING

In producing gas and oil, more and more importance is being placed upon most efficient recovery (MER) performance of the producing wells. Generally, some sort of test must be made to determine the performance of the well.

Potential Test

The most frequently conducted well test is the potential test, which is a measurement of the largest amount of oil and gas a well will produce in a 24-hour period under certain fixed conditions. The produced oil is measured in an automatically controlled production and test unit or by wire-line measurement once the oil is out of the ground and in the lease tank.

This kind of test is normally made on each newly completed well and often during its production life. The information obtained from the test is generally required by the state regulatory group, which assigns a producing allowable that must be followed by the well operator. At later intervals, similar tests are made and producing allowables are adjusted according to the results of the tests. Often, these tests help establish proper production practices (covered in a later section).

Bottom-Hole Pressure Test

This is a measure of the reservoir pressure of the well at a specific depth or at midpoint in the producing interval. This test measures the pressure of the zone in which the well is completed. In making this test, a specially designed pressure gauge is lowered into the well by means of a wire line. The pressure at the selected depth is recorded by the gauge, and the gauge is then pulled to the surface and taken from the well.

There are several variations to this type of test, such as the flowing bottom-hole pressure test, which is a measurement taken while the well continues to flow. A shut-in bottom-hole pressure test is a measurement taken after the well has been shut in (closed) for a specified length of time. These tests also give information about the fluid levels in the well. Other bottom-hole pressure tests furnish valuable information about the decline or depletion of the zone the well is producing.

Productivity Tests

These tests are made on both oil and gas wells and include the potential test and the bottom-hole pressure test. This reading determines the effects of different flow rates on the pressure within the producing zone of the well and thereby establishes certain physical characteristics of the reservoir. In this manner, the maximum potential rate of flow can be calculated without risking possible damage to the well.

In the procedure, the closed-in bottom-hole pressure of the well is first measured. Then the well is opened and produced at several stabilized rates of flow. At each rate of flow, the flowing bottom-hole pressure is measured. These data provide an estimate of the maximum flow expected from the well.

Fluid-Level Determination

Fluid-level determination is a test most commonly performed on wells that will not flow and must be made to produce by pumping or other means of artificial lift. To help select the proper equipment, the standing fluid level in the well is determined. First, a small explosion is created at the wellhead. The sound is deflected by tubing couplings. By counting these deflections (echoes), the fluid level in the well can be determined.

Bottom-Hole Temperature Survey

This test is normally made along with the bottom-hole pressure test. It determines the temperature of the well at the bottom of the hole or at some point above the bottom. First, a specially designed recording thermometer is lowered into the well on a wire line. After the thermometer is extracted, the temperature of the well at the desired depth is read from the instrument. These data and the bottom-hole pressure calculations are used to solve problems about the nature of the oil or gas. Temperature tests are sometimes helpful in locating leaks in the pipe above the producing zone. They are also used to determine whether gas-lift valves are operating, the location of top and bottom cement in newly cemented wells, and the injection interval in injection wells.

PRIMARY RECOVERY

So far discussed have been the unusual methods of primary recovery. That is, the initial production of hydrocarbons from a well or field. During this phase of operations it can be expected that approximately 25 percent of the oil in place in the reservoir can be recovered. Do not get the impression, however, that pumps run around the clock sucking every available drop of oil as rapidly as possible. To do so would, in effect, be a very wasteful practice.

The recovery of wanted petroleum fluids depends on the ability of those fluids to flow through the formation to the well, and this of course is in turn dependent on the porosity and permeability of the formation itself. The idea, then, is to maintain a steady rate of flow (depletion) that will prolong the useful primary life of the reservoir as much as possible.

The term *pumping off* as used in Table 6–1 means that usually there will only be so much oil available at a given well. For example, a pump may be timed to run for four hours at a time. During this period, the oil at the base of the well will have been exhausted. A rest period for the pump will then follow while a new supply of oil flows through the formation to the well. At a preset time, the pump will restart for another pumping cycle. If the pump were allowed to run continuously, the flow of oil through the formation could be stretched so thin that salt water could intrude and isolate pockets of otherwise recoverable oil; thus, much of the total oil in place would remain wasted in the ground.

How long to operate each pumping cycle will be calculated by the reservoir engineer, who will take into consideration the total estimated oil in place, the porosity and permeability of the formation, the bore and stroke of the pump, and other factors. Such recovery methods were largely unknown to, or ignored by, early operators who worked hard to produce as much as they could as fast as they could. Too, at various times government regulations and pro rata agreements set arbitrary production figures that were not based on sound scientific or engineering principles. Many fields, therefore, were pumped dry and abandoned, when in reality they still contain a great deal of oil. On the other hand, because no excessive pumping methods were ever used, many wells are still maintaining a satisfactory daily level of production after years of operation.

ENHANCED OIL RECOVERY (EOR)

After the primary recovery period of a reservoir is over, the area may not be abandoned as was done in the past, but efforts commence to obtain all the petroleum possible through the techniques of secondary and even tertiary recovery.

Secondary Recovery

It is estimated that more than 100 billion barrels of oil may still remain in depleted reservoirs in the United States alone. Secondary and tertiary-recovery techniques can increase or maintain the present levels of production for many years to come.

Secondary-recovery processes generally use the injection of water or natural gas into the production reservoir to replace or assist the natural reservoir drive or primary production. One method is to drill a well in the center of a four-well producing location and inject water under pressure into the reservoir (Fig. 6–8). The injected water forces some of the crude oil away from the injection well and toward the producing wells (water flood).

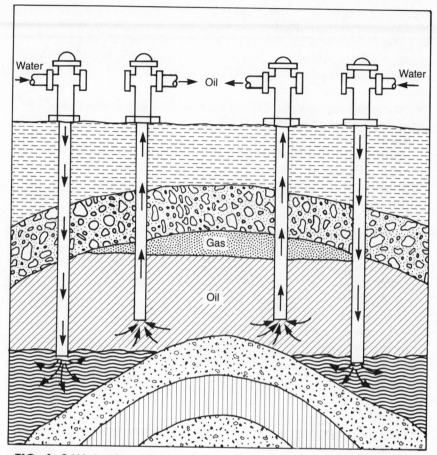

FIG. 6–8 Water-flood technique.

When natural gas was abundant and cheap, it was also injected under pressure into the reservoir. However, U.S. natural gas prices have caused a decline in its use for injection.

Conventional secondary-recovery methods do not completely deplete the reservoir. At or near the end of their effective life, tertiary recovery may be used to recover more oil from the reservoir.

Tertiary Recovery

Tertiary-recovery methods go ahead where secondary-recovery methods leave off. They are usually divided into three major categories: thermal, chemical, and miscible displacement.

Thermal processes include steam stimulation, steam flooding, and in-situ combustion. Steam stimulation involves injecting steam into a producing well for 2–3 weeks. The well is shut in so the heat can dissipate and transfer to the oil. As the oil warms up, it begins to flow more easily, increasing production. Steam flooding is similar to water flooding. Input wells are located in a pattern for injection while the oil is produced from adjacent wells. The injected steam or hot water forms a saturated zone around the input well. As the steam moves away, its temperature warms the oil, moves it, and leaves steam to displace the vacant pores. In-situ combustion (also called fire flooding) involves igniting an air-injection well. The combustion front moves away from the well, heating the oil so it moves more easily and pushing it toward a producing well. This method can be transferred back and forth between several wells until the reservoir is better depleted (Fig. 6–9).

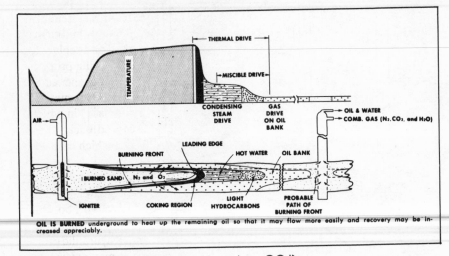

OIL IS BURNED underground to heat up the remaining oil so that it may flow more easily and recovery may be increased appreciably.

FIG. 6–9 In-situ combustion (courtesy OGJ).

Chemical processes include surfactant-polymer injection, polymer flooding, and caustic flooding. In the surfactant-polymer process, a surfactant slug is injected, then a polymer mobility slug is injected as a buffer. This process is repeated, providing a cleansing agent and then a moving agent. The polymer flooding process adds a thickening agent to the water in a waterflood system. The polymer reduces the volume of water required and increases sweep efficiency. Water takes a path of least resistance, i.e., it doesn't spread out but tends to form rivers or channels. The polymer enhances water's sweep. Caustic or alkaline flooding changes the pH of water, making it more acidic. This helps open passages in the pores and increases production.

Miscible-displacement processes include miscible hydrocarbon displacement, carbon-dioxide injection, and inert gas injection. Miscible hydrocarbon displacement uses a solvent that is injected and followed by an injection of a liquid or gas moving agent. The carbon-dioxide injection process mixes CO_2 with the crude. After a proper period, the CO_2 forms a miscible front that pushes the oil to the production well. Inert gas injection is usable for only a select few wells. An inert or almost totally unreactive gas is added to push the oil through the rock.

All of these processes are still in the beginning stages of experiment. Presently, they each apply to specific kinds of reservoirs, and they do not cleanse the formation 100%. New research and methods will be needed before we can glean every drop of oil from beneath the earth.

WELL SERVICING

As with almost any device that is man-made, wells require periodic servicing and workover. This may be no more than minor maintenance on pumping equipment to major overhaul and repair. Pumps and tubes may become packed with sand and thus have to be pulled and replaced. Casing is subject to corrosion, which is a major problem facing producers, and it may have to be pulled and replaced. Aciding and additional fracturing may be needed to stimulate production. Or it may be decided to produce from a level closer to the earth's surface, in which case the lower portion of the well will have to be plugged back. Conversely, production from lower levels may be desired, in which case the well is drilled deeper.

If the main bore is blocked by unretrievable broken tools, then a sidetrack well may have to be drilled to bypass the blockage to get to the deeper zones (Fig. 6–10). To control casing leaks and to shut off bottom water, cement may have to be added by squeeze cementing, where the cement is mixed into a slurry and forced into place under hydraulic pressure using a tool called a squeeze. Finally, the service may acidize or hydraulically fracture the well to reopen the pore spaces around the well bore.

FIELD PROCESSING

Virtually all petroleum produced from a reservoir requires processing. Petroleum is a complex mixture of many different compounds of hydrogen and carbon, all with different densities, vapor pressures, and other physical characteristics. Many well streams are a turbulent, high-velocity mixture of gases, oil, free salt water, salt-water vapor, solids, and impurities. As the stream flows, it undergoes continuous pressure and temperature reduction because it is leaving a high-pressure reservoir that is hotter than the earth's surface it is coming in contact with.

Often, petroleum mixtures are very complex and difficult to separate easily and efficiently. Hazards of fire, loss of vapor, and evaporation must be eliminated or controlled by the equipment used in a separation process.

Once the oil or gas flows out of the wellhead, it enters a flow line and is carried to the *header*. The header is the junction connecting all flow-

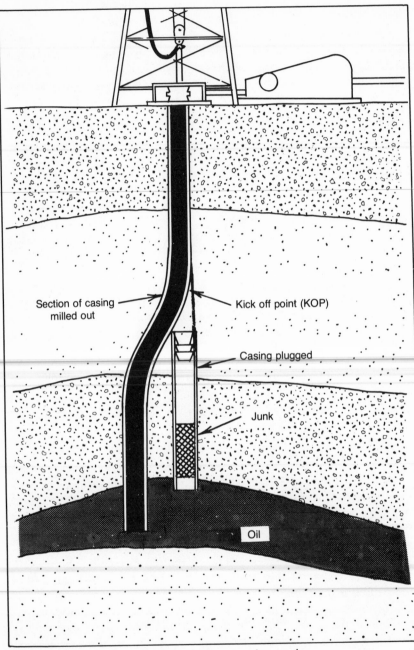

FIG. 6–10 Sidetrack drilling from a workover rig.

lines in a given area, and it consists of different kinds of valves and fittings. From here, the oil and gas enter the separator.

Separators

Closed vessels called *separators* remove natural gas from oil and water and oil from water. The simplest form of separator is a closed tank in which the force of gravity separates the gas, oil, and water. The force of gravity has a greater pull on water because it is heavier, and it settles to the bottom of the tank. Gas is the lightest and moves to the top of the tank and on to a dehydrator, leaving oil floating on top of the water. Thus separation is caused by their *specific gravity* differences (Fig. 6–11). Gravity separation takes time, which may not be a factor for some low-production wells. High-production wells often need special equipment for efficient separation.

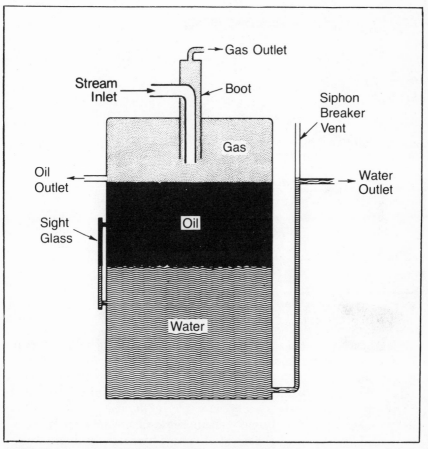

FIG. 6–11 Oil and gas gravitational separator.

Separators are usually classified in two ways: the shape or position of the vessel and the different number of fluids segregated. The three common vessel shapes are vertical, horizontal, and spherical. The number of fluids to be segregated is normally either two or three. The separator is referred to as a two-phase type if only gas and liquid are segregated. The three-phase type segregates gas, oil, and water.

Vertical separators. Vertical separators may be either two-phase or three-phase, depending upon the field requirements (Fig. 6–12). Vertical separators are often used on low to intermediate gas-oil ratio wellstreams. They are also used where large slugs of liquid are expected and liquid level control is not critical.

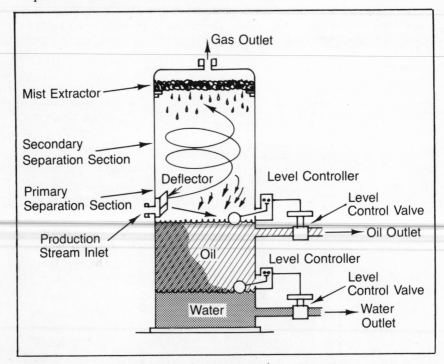

FIG. 6–12 Vertical three-phase separator.

Vertical separators are normally equipped with an inlet diverter that causes the incoming fluids to swirl in a centrifugal motion, providing the proper momentum reduction that allows the gas to escape. The liquid falls to the liquid accumulator at the bottom and the gas rises to the top. Separation is not complete yet because some of the small liquid particles are swept upward with the gas stream. These small liquid particles are separated by a baffle arrangement of knitted-wire mesh pad or a vane-type *mist extractor* positioned near the top.

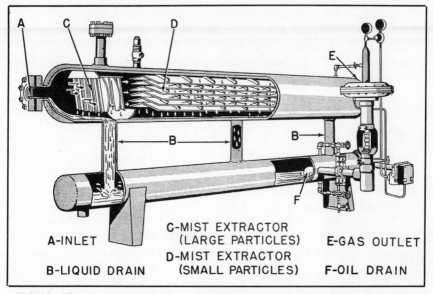

A—INLET

B—LIQUID DRAIN

C—MIST EXTRACTOR
(LARGE PARTICLES)

D—MIST EXTRACTOR
(SMALL PARTICLES)

E—GAS OUTLET

F—OIL DRAIN

FIG. 6–13 A double-tube horizontal separator. Oil drops down the bottom and goes to the stock tanks. Gas leaves through the gas outlet.

A vertical separator is much taller than the horizontal or spherical but occupies less foundation space. It handles large quantities of sand or other solid impurities and is easier to clean out. One disadvantage is that it is larger than the others of the same capacity and costs more.

Horizontal separators. Horizontal separators may be two-phase or three-phase and have a greater gas-liquid interface area that permits much larger gas velocities than other types (Fig. 6–13). They are generally used for high gas-oil ratio wellstreams and are more efficient and economical for processing large volumes of gas. They are also used for handling foaming crudes or for liquid-from-liquid separations.

The horizontal separator is easier to transport and connect with piping and fittings. It requires less overall space, is more adaptable for skid mounting, and can be stacked for stage separation, such as on offshore platforms.

The interface area of the horizontal separator consists of a large, long, baffled, vane-type gas separation section. Gas flows horizontally and at the same time slants toward the liquid surface. The moist gas flows in the baffle surfaces and forms a liquid film that is drained away to the liquid section of the separator.

Horizontal separators of the two-tube, or double-barrel, design are used in some locations where a large amount of free liquid is in the wellstream. It has all the advantages of a conventional horizontal separator plus a higher liquid capacity.

Spherical separators. Spherical separators are more compact than the vertical or horizontal separators and make maximum use of all the known methods of oil and gas segregation. They usually cost less than the other two. The disadvantages are that the surge capacity is limited and it is not as economical for larger gas capacities.

Oil Treating

After crude oil leaves the field separator, it will often be emulsified and require further segregation to make it ready for sale and transporting. Saleable quality usually means containing a maximum of 1% BS&W (basic sediment and water).

Emulsions are mixtures of oil and salt water that appear like muddy-brown foam. They usually do not exist in the production formation but are caused by agitation as the wellstream flows through all the piping and equipment.

Close observation will reveal a great number of very small spheres of water mixed through the crude oil. These tiny water spheres are surrounded by a tough film caused by the difference in *surface tension* of the water and oil. The film must be destroyed in order for the water to drop out.

Certain actions tend to weaken the film so that the droplets can come together to form larger drops that will settle to the bottom. Heat weakens the film as do certain "demulsifier" chemicals. After the emulsion is broken, settling time is required to complete the treating method. Some emulsions can be destroyed with chemicals and settling and others can be destroyed by heat and settling time. The more difficult emulsions require chemicals and heat followed by settling time.

Chemical treatment is normally done by chemical injection equipment that applies the chemical downhole, in a flowline, or by batch treating.

Treating emulsions with heat requires special independent equipment usually called a *heater treater*. These are usually either constructed as vertical vessels or horizontal vessels. Flames at the bottom of the equipment (usually from natural gas) heat the emulsion. The oil rises to the top and is carried to settling tanks. The water sinks to the bottom and flows through pipelines back to the well where it can be used in water-drive secondary recovery.

Another method of treating is done with electrostatic treaters. They are similar to horizontal heater-treaters except that high-voltage electric grids are added. Electricity is sometimes an effective means of breaking emulsions.

Most field separation and processing equipment will operate under pressure. The equipment is usually constructed of steel and all seams are welded. It is built according to strict pressure vessel specifications. Pressure relief valves or other suitable protection is provided for safety. Because steel is used in this equipment, severe corrosion is encountered. To combat this, corrosion-resistant materials are used whenever possible or corrosion protection is provided.

CHAPTER 7

Natural Gas

Natural gas is always present with crude oil from a reservoir. The quantity and quality usually will vary for each reservoir. Natural gas is a homogeneous fluid of low density and low viscosity. It is classified as a fluid, and fluids include both liquids and gases. Unlike liquids, however, gases have neither definite shape nor definite volume, i.e., gas will expand to fill its container.

Natural gas is one form of energy that is basically a mixture of hydrocarbon (hydrogen and carbon) with some impurities. Two of the most unwanted impurities are sand and salt water. These impurities should be removed as early as possible from the gas stream. Usually they can be removed with separators and scrubbers located near the wellhead. (A *scrubber* is a vertically mounted vessel with baffles and screens inside and a collecting area for impurities.) Some of the other impurities, such as small heavier hydrocarbons, carbon dioxide, and nitrogen, remain in the gas until it is gathered at a central plant then removed in single units at a lower cost.

Usually the hydrocarbon gases found in natural gas are methane, ethane, propane, butanes, pentanes, and small amounts of heptanes, octanes, hexanes, and heavier gases. The propane and heavier fractions are removed and later processed for their value as gasoline-blending stock and chemical plant raw feedstock.

Methane and ethane are the most abundant mixtures found in natural gas that have a real value as fuel. These increase the BTU (British thermal unit) value of the gas. In other words, the more methane and ethane in gas, the higher the heating value and the greater the quality of the gas. The normal components of a typical natural gas at the wellhead are shown in Table 7–1.

TYPES OF NATURAL GAS

The four general types of natural gas are wet, dry, sweet, and sour. Wet gas contains some of the heavier hydrocarbon molecules and water vapor in the reservoir. When the gas reaches the surface, a certain amount of liquid is formed. The water is of no value; however, the liquid

TABLE 7-1 Typical Natural Gas Components (after McCain)

Hydrocarbon	Amount, %
Methane	70—98
Ethane	1—10
Propane	trace—5
Butane	trace—2
Pentane	trace—1
Hexane	trace—½
Heptane +	none—trace
Nonhydrocarbon	
Nitrogen	trace—15
Carbon dioxide	trace—1
Hydrogen sulfide	trace occasionally
Helium	trace—5

hydrocarbon molecules are of value for processing operations. If the stream does not contain enough of the heavier hydrocarbon molecules to form a liquid at surface conditions, it is dry gas. Sweet gas has a very low concentration of sulfur compounds, particularly H_2S. Sour gas contains excessive sulfur compounds and has an offensive odor. It can be harmful to breathe. In fact, excessive amounts of H_2S are fatal.

TYPES OF WELLS

There are three general types of production wells that produce natural gas: oil wells, gas wells, and condensate wells (Fig. 7-1). In some fields, various types of pumping units are visible at the wellhead with stock tanks and separators nearby. These are the typical oil wells that were drilled primarily for the crude oil that they produce. Some of the wells will also produce enough natural gas to make processing and handling profitable.

Gas wells are obviously different than oil wells. One of the differences is at the wellhead where only pipe connections, valves, and gauges are visible. These configurations are commonly referred to as the *Christmas tree*. Since the gas flows from the reservoir through pipe to the closed vessels above ground, no pumping unit is required. Some gas wells produce some crude oil and require stock tanks to contain it. Also, some oil wells flow and do not require pumping units. Gas wells are normally drilled for the natural gas they produce.

Condensate wells are usually flowing wells with the Christmas tree connected to the casing at the surface. They produce both natural gas and condensate. Condensate is a liquid hydrocarbon that lies in a range between gas and oil. It is separated from the gas by cooling and various other means and then is stored in vessels for future use.

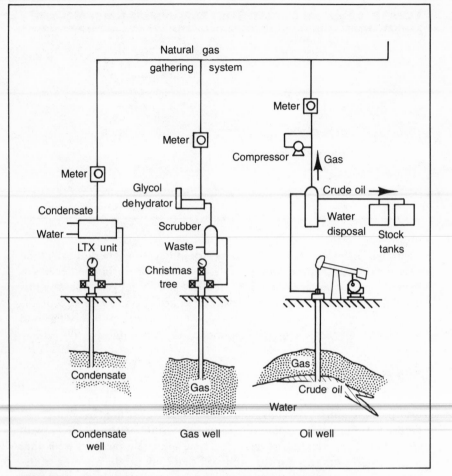

FIG. 7–1 Oil, gas, and condensate well schematics.

PHYSICAL PROPERTIES OF NATURAL GAS

Each production stream of natural gas may vary in composition and in the relative amounts of compounds; thus, their physical properties will vary, also. In order to predict the behavior of a particular stream of gas for processing purposes, it is important to know its physical properties. One method of doing this is to take samples of each stream and have laboratory analysis made. This requires the use of a specially designed gas bottle for taking samples.

Once the natural gas composition is known, the different physical properties of each pure component can be determined. Physical properties are usually termed either intensive or extensive. *Intensive properties* are independent of the quantity of material present and include density, dew point, specific gravity, critical temperature, and critical pressure.

Extensive properties are determined by the total quantity of matter present, such as volume and mass.

Since gas is a fluid, it is important to examine the physical properties of fluid. A fluid is a substance that flows. A substance could be in a liquid form under certain conditions, then change to a vapor or gas under the other conditions, or vice versa. For example natural gas becomes liquefied natural gas (LNG) when subjected to a very low temperature.

Every fluid that includes liquids, gases, and vapors has many properties that can make it different from the other fluids. Some of these properties could be temperature, pressure, gravity, critical temperature, critical pressure, bubble point, dew point, boiling point, hydrate point, latent heat of vaporization, molecular weight, density, miscibility, and phase.

Temperature is one of the very important properties of fluids. Each fluid has a property that is called *critical temperature*. When fluid is at a temperature higher than its critical temperature, it will not be in a liquid form. When a fluid is at a temperature lower than its critical temperature, then it will be either a liquid or vapor, depending on the pressure acting on it.

Pressure is the weight or force that a liquid applies on a defined area, usually one square inch of area. When the defined area is one square inch, then pressure measurements are expressed in pounds per square inch (psi).

Each fluid has a property called *critical pressure*. A liquid mixture must be below its critical pressure to be separated.

When two liquids will not mix, they are called *immiscible,* such as gasoline and water. If gasoline and water are put into the same tank, almost all of the water could be drained from the tank, leaving the gasoline. When two liquids do mix without separating, they are *miscible,* such as gasoline and butane. If gasoline and butane are put into the same tank, the gasoline, being heavier, can be drained off leaving only the butane.

The gases that make up natural gas have different masses and weights. *Weight* is generally synonymous with *mass,* so the weight of a material rather than the mass is normally used as a measure of quality. At sea level, mass and weight are equal.

The *structure* of gas and its *molecular weight* are related. The formula of a gas indicates the relative numbers and kinds of atoms that unite to form the gas molecule. The methane formula CH_4, for example, indicates that carbon and hydrogen are present in the compound in a 1:4 ratio. By taking the atomic weight of carbon and adding to it four times the atomic weight of hydrogen, the molecular weight of 16:043 can be obtained.

Density is the weight (or mass) per unit volume. The density of natural gas is usually expressed as the weight in pounds per cubic foot (lb/cu ft). Usually, the volume are at the standard condition of 60 °F and 14.7 psia. Air has a normal density of 0.0763 lb/cu ft, and natural gas will be lighter than air. Specific gravity is the ratio of a gas density to the density of air at the same conditions of temperature and pressure.

Latent heat of vaporization is the heat necessary to change a liquid to a gas.

Boiling point is when a liquid will boil whenever the vapor pressure of a liquid is equal to the pressure being exerted on it. Since the vapor pressure of a liquid changes with temperature, any liquid has many diffrent boiling points, depending on the pressure being exerted on the liquid.

The temperature at which the hydrocarbon mixture begins to vaporize at a given pressure is the bubble point. It is the temperature at a given pressure at which the first bubble would form.

Dew point is the temperature at a given pressure at which the first drop of liquid forms in the gas system. Or it is often referred to as the point at which condensation takes place.

Specific heat is defined for practical purpose as the number of BTUs needed to raise the temperature of one pound of material 1 °F.

Hydrates are created by a reaction of natural gas with water. They are solid or semisolid compounds that form ice-like crystals. The actual compostion of natural gas and pressure determines the temperature at which hydrates will form when it begins to cool. When hydrates form in gas lines, they coat the insides of the pipe and restrict flow.

A *phase* can be defined as any homogeneous and distinct part of a system that can be physically separated from other parts of the system by distinct boundaries. Gas processing is concerned with the two phases, gas and liquid. Thus, vapor and gas are synonymous.

FIELD PROCESSING

Usually all natural gas taken from a producing well requires some processing. Generally, natural gas is processed near the wellhead to remove unwanted impurities and harmful hydrates. Field processing equipment will vary widely between locations. It is usually constructed of steel and will operate under pressure. Some of the more common pieces of equipment will include conventional separators, low-temperature separators (LTX), heaters, dehydrators, and scrubbers.

Low-temperature separators are often used with high-pressure gas wells that produce very light crude oil or a condensate. The process often uses the effect of expanding high-pressure gas across a special choke to obtain a cooling effect. The LTX system separates the water and hydrocarbon condensate (propanes, butane, and light gasoline) from the inlet wellstream. It recovers more liquids from the gas than can be recovered with the normal-type separators.

Heaters are often used to heat the gas stream to above-ground temperature to control hydrate formation. Heaters are relatively simple and require little maintenance or attention. They use natural gas for fuel and are less expensive than dehydration units. They are not as efficient as dehydrators, and often heating must be repeated at each point in the system where hydrate formation is possible.

The two general types of heaters are the flow-line heater and the indirect heater. The *flow-line heater* operates by directly heating the pipes with a gas flame in an enclosed chamber. *Indirect heating* can be done by passing the gas lines through tanks of water heated by fired heaters. The indirect heater is more commonly used than the flow-line heater (Fig. 7–2).

The term *dehydration* means removing water from a substance; in this case the substance is natural gas. The process used to remove water

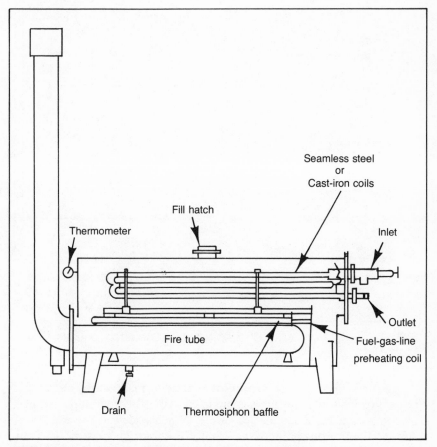

FIG. 7–2 Cutaway view of an indirect heater.

from gas is either absorption or adsorption. *Absorption* means the water vapor is sucked up or taken by an agent, such as glycol. This requires a reaction. Adsorption means the vapor is collected in condensed form on the surface. It requires no chemical reaction.

One general type of dehydration is the liquid-desiccant dehydrator using glycol as the absorption agent. Glycol has an affinity for water (Fig. 7–3).

The other type of dehydration is the solid-desiccant dehydrator using activated alumina or a silica-gel-type granular material. Water is retained on the surface of the particles of a solid material as the wet gas is passed through. Both types of dehydrators require a regeneration process to remove the water from the glycol or solid-desiccant (Fig. 7–4).

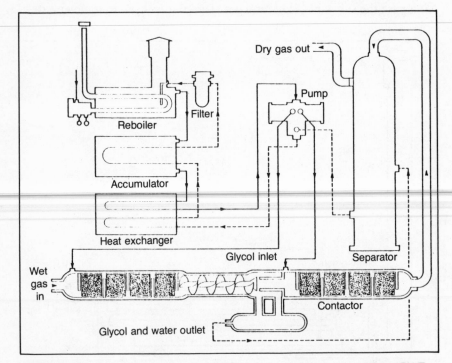

FIG. 7–3 Flow diagram of a horizontal contractor (courtesy BWT-Moore).

Another method used to prevent hydrate formation is by injecting a stream of inhibitor into the gas piping system. The inhibitors could be glycol, methanol, ammonia, or brines. When added to the water-wet stream, the freezing point of the water is lowered. This system does not remove water from the gas; it only helps prevent hydrate formation.

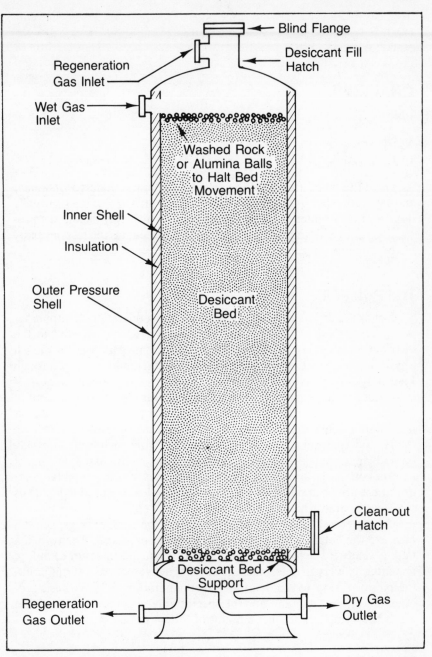

FIG. 7-4 Adsorber tower.

CHAPTER 8

Storage

Once the fluids are out of the ground, they must undergo treatment and separation before they are sent on to market. The first step is usually done at a tank battery located on the lease, which may handle the production of several wells. At this point, the petroleum is the responsibility of the pumper, or gauger.

THE PUMPER

Just as the driller is the key person in drilling operations, the pumper is the key person in transferring the oil and gas from the well to the distribution system or initial purchaser. The job may not call for rugged physical characteristics, but the pumper or gauger must have a thorough knowledge of the job and equipment, be dependable, and be honest. It is the duty of the pumper to produce and measure the proper amount of gas and oil from the well and to see that the owner gets proper credit for the amount delivered.

The pumper must control the production from each well. This may be set by allowables—state regulations set each month based on current market conditions and efficient production rates for the particular field; or if there are no applicable regulations, then whatever production goals the lease operator has established.

During each 24-hour period of operation, the pumper or gauger will measure the volume of gas, oil, and salt water produced. Then when a tank is ready for delivery, the tank must be gauged (measured) before and after delivery to determine the amount delivered. Also at this point, the oil must be tested to determine its temperature, BS&W (basic sediment and water) content, and the specific gravity of the oil, factors that affect the value and the net amount of the oil passed on.

THE STORAGE SYSTEM

The fluids from the well are piped to the tank battery, which usually consists of a gas-oil separator that separates these two fluids, and from

124

which the oil passes on to a gun-barrel tank, an oil treater that removes the water and sediment, and finally into storage tanks to await shipment (Fig. 8–1).

FIG. 8–1 A typical lease tank battery (courtesy Ketal Oil Producing Co.)

Since more than one well may be feeding into the battery, gas meters and test gas-oil separators may also be provided for measuring production from each well. Also, an oil meter may be used for well testing so that the tank measurement of a well being tested can be combined with the oil from the other wells. Such testing may be required by the state. In any event, it is needed by the operator of the lease to maintain efficient operation of the wells and the reservoir.

Tank Construction

Most tanks are made of either bolted together or welded steel plates, although in areas where corrosive crude oil is produced, wooden gun barrel and stock tanks may be used. Fiberglass tanks are also coming into use. Stock tanks will have a bottom drain outlet for removal of the BS&W, and there may also be an access or clean plate. This is a provision made near the bottom of the tank for a workman to enter and periodically clean out paraffin and sediment that have accumulated and will not drain off. Proper safety precautions must, of course, be observed during such operations.

The gun-barrel tank will usually be fitted with a salt-water siphon line, and filling and equalizing lines will run from it to the stock tanks.

The oil enters the stock tanks at the top. An inlet valve is installed so that once the tank is filled and ready to deliver, no more oil will enter. A gas vent line is also provided across the top of the stock tanks.

The delivery connection is usually placed one foot above the bottom of the tank; this leaves space for the collection of the BS&W. There are provisions made for sealing the outlet valve between deliveries. On the top of the tank is an opening called the *thief hatch,* through which measurements can be made and samples extracted (Fig. 8–2). An earthen dike may be built around the battery as a safety precaution in the event of fire or tank rupture.

FIG. 8–2   Preparing to thief a tank at the thief hatch.

The size and number of tanks in the lease battery will depend upon the daily production and the frequency of deliveries, which may be directly into a pipeline or by means of tank cars or trucks. The usual capacity of the battery is equal to 3 to 7 times the daily production of the wells feeding the battery. Normally, two or more stock tanks are provided so that while oil is being delivered from one tank, another may be filling.

The capacity of welded tanks may run from 90 to 3,000 barrels; bolted tanks hold from 100 to 10,000, wooden from 200 to 1,000 barrels.

Strapping

Before a tank is put into service for the first time, it must be *strapped,* meaning its dimensions must be measured and the volume of oil it can hold calculated. These data are used to construct a *tank table* or chart showing the amount of oil in barrels at given intervals, usually ¼ inch apart, from the bottom to the top of the tank. This process is usually carried out by a third or neutral party.

OPERATION

As the fluids pass through the separator, the gas is measured as it flows through an orifice meter and the oil is directed to one of the tanks. When one stock tank is filled, the flow is then directed to another. The pumper or gauger has made periodic measurement of the contents of the tank, but the final gauging before running or delivery is made by the *pipeline gauger* or transport driver picking up the oil. He also makes the final gauge when the tank is empty and seals the outlet valve so the tank may be refilled. The pumper watches the process and verifies the measurements and tests. This is important, for it is at this point that the custody or ownership of the oil passes from the first owner, the lease holder, to the transportation company. The pumper or lease gauger reperesents the next owner or custodian.

Ever since the impact of the energy crisis, thefts of crude oil from lease tank batteries have been on the increase and new security precautions are now being taken to protect the owners of the oil.

MEASUREMENT AND TESTING

The lease pumper will usually also take care of the gas metering equipment to make sure that it is operating properly and that gas deliveries are being recorded. This will include the changing of the recording charts and pens and forwarding the charts to the lease holder's office so that the amount delivered is calculated and the recipient is billed.

Salt water produced with the oil must be removed from the tanks. On the gun-barrel tank this may be handled automatically by a siphon but must be done manually by means of the drain valves on the stock tanks. The daily salt-water production is measured by gauging the tank before and after the salt water is removed. In the past, salt water was disposed of by running it into the nearest creek. Now, however, environmental concerns have brought about regulations providing for salt-water injection or purification.

Some oil in the lower level of a tank may have become emulsified with hydrocarbons. This is transferred to a treater so that salt water and the sediment can be removed. Additional oil will be added to the tank to top it off or replace that sent through the treater so that a full (or almost full) tank can be delivered. The treated oil will be sent to a tank not yet ready for delivery.

Testing Procedures

Testing and sampling are usually done through the opening in the top of the tank called the thief hatch, although samples may also be obtained

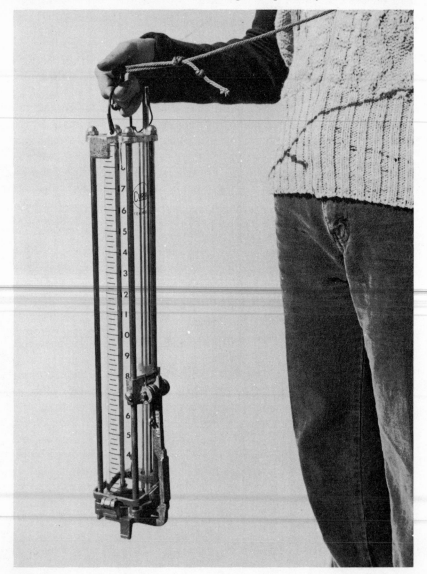

FIG. 8-3 Tank thief.

through the sidewall of the tank if appropriate taps have been provided for that purpose. Several tools have been devised to make the tests and measurements necessary before sale or transfer of the oil. The primary tool lowered through the hatch is the *thief,* and its use is known as *thiefing* the tank (Fig. 8–3). The tool is lowered first to the top third, then to the middle third, and finally to the bottom third. Then an average of the three measurements is calculated to find the API gravity of the oil.

Other gauger's tools include a hydrometer and graduate for measuring the API gravity of the oil, a thermometer for measuring the temperature of the oil, tank-gauging line for measuring the height of the oil in the tanks, and a centrifuge for measuring BS&W content.

The price paid by the purchaser for a tank of oil will depend on the amount (measured in 42-gallon barrels), the weight of the crude as expressed in degrees API gravity, and the amount of BS&W in the oil. In the early days, lighter crudes which ran toward the natural gasolines and aromatics were the most desirable and therefore commanded a higher price. This was because the majority of the oil purchased was for conversion to motor fuel. With the advent of the petrochemical industry, however, the heavier crudes have grown in demand since they are necessary for the manufacture of such items as plastics and synthetic rubber. The price per barrel the producer receives is still dependent on the specific gravity of the crude despite this change in the market.

Specific Gravity

By the nature of the field in which it is located, one lease may produce a very "light" crude while the oil from another lease in another field may be thick, viscous, and odiferous. In other words, not all crude oil is alike. The first step in scientifically measuring the difference between different samples of crude oil is to determine the *specific gravity* of the sample at hand.

Specific gravity is the ratio between the weight of a unit volume of a substance compared with the weight of an equal volume of some other substance taken as a standard. For liquids, the standard is usually water, which is assigned a value of 1. For gases, the standard is usually air, also assigned a value of 1. (This is assuming a temperature of 60°F.) Since the temperature at the tank battery is not always going to be 60°, temperature conversion tables will have to be used to determine the correct specific gravity.

API Gravity

The American Petroleum Institute has adopted a standard method of expressing the gravity, or unit weight, of petroleum products. This is an arbitrary method that had its beginnings in the chemical industry long before it was applied to liquid hydrocarbons or before much was really known about the relationships between the weights of substances. API gravity may be determined by placing a sample of the oil in a hydrometer and reading the value directly off the scale or by a fairly simple computation.

The first step in computing the API gravity of a sample is to determine its specific gravity. The API formula is:

$$\text{API gravity} = \frac{141.5}{\text{specific gravity at 60 °F}} - 131.5$$

Thus if a certain sample had a specific gravity of 0.82, its API gravity would be:

$$\frac{141.5}{0.82} = 172.6 - 131.5 = 41.1° \text{ API}$$

Temperature

The temperature of the oil in the tank will usually be close to that of the outside air unless it has just been pumped, in which case it will be warmer due to the warmer temperature of the earth. It may also show an elevated temperature if it has just been run through a heat treater to remove BS&W. The usual method is to lower a thermometer on a line through the thief hatch and match the reading, if it is other than 60°F., to a conversion table. Such correction tables are published by the American Society for Testing Materials (ASTM).

BS&W

Samples of the oil are also taken to measure their content of basic sediment and water. A sample is removed from the tank by means of a thief or *sidewall cock* (spigot or faucet in the wall of the tank) and placed in a centrifuge where it is shaken out—that is, the different elements of the sample separate and the results are read directly on a scale.

STANDARDS

The purchaser of the oil certainly is going to test it before delivery is made; therefore, the producer should be doing his own testing to determine if the oil is going to be acceptable. Thus, sampling and testing are carried out in accordance with standards set up and adopted by the API and the ASTM. Samples (seldom, if ever, is just one used) may be drawn using a thief, bottles, or sidewall taps or cocks (Fig. 8–4). Thus, samples are taken from several different levels of a tank. The thief itself is a round tube, usually 15 inches long, that is suspended from a chain connected to its top. At the bottom is a spring-operated sliding valve that can be actuated to seal the bottom end. It is lowered into the tank with

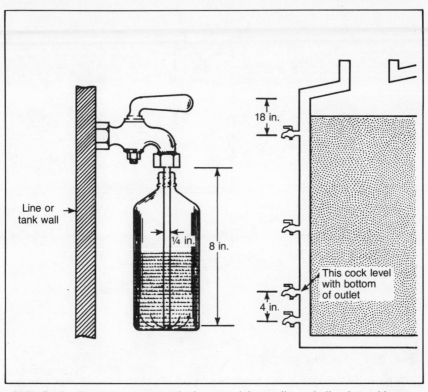

FIG. 8–4 Drawing a sample from a side-wall cock (top) and location of the taps (bottom).

the valve open so the oil passes freely through the tube until the desired depth is reached. At that point, the spring is released and the valve closes, capturing a sample of oil. The thief is then brought out.

A one-quart metal bottle may also be used. The bottle is weighted so that it will sink and has a cork stopper. A line is attached to the stopper so that it may be removed at the desired depth and the bottle filled at that point (Fig. 8–5).

There are several different types of samples that may be taken and an agreement should be reached by all concerned as to which type will be used.

An *average sample* consists of proportionate parts from all sections of the tank.

An *all-levels sample* is obtained by submerging a stoppered beaker or bottle to a point as near as possible to the drawoff level, then opening the sampler and raising it at such a rate that it is about three-quarters full (maximum 85 percent) as it emerges from the liquid. An all-levels sample is not necessarily an average sample because the tank volume may

FIG. 8-5 Beaker sampler.

not be proportional to the depth and because the operator may not be able to raise the sampler at the variable rate required for proportional filling. The rate of filling is proportional to the square root of the depth of the immersion.

A *running sample* is one obtained by lowering an unstoppered beaker or bottle from the top of the oil to the level of the bottom of the outlet connection or swing line and returning it to the top of the oil at a uniform rate of speed so that the beaker or bottle is about three-quarters full when withdrawn from the oil.

An *upper sample* is a spot sample taken at the midpoint of the upper third of the tank contents (Fig. 8–6).

A *middle sample* is a spot sample obtained from the middle of the tank contents.

A *lower sample* is a spot sample obtained at the level of a fixed tank outlet or the swing-line outlet.

A *clearance sample* is a spot sample taken 4 inches (100 millimeters) below the level of the tank outlet.

A *bottom sample* is obtained from the material on the bottom surface of the tank, container, or line at its lowest point.

A *drain sample* is obtained from the draw-off or discharge valve.

A *water and sediment sample* is obtained with a thief to determine the amount of material at the bottom of the tank that cannot be sold.

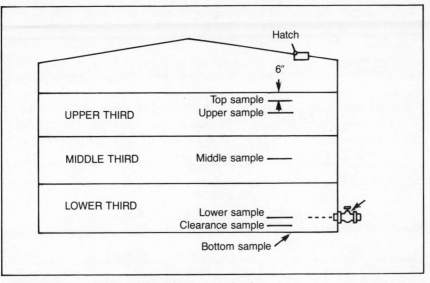

FIG. 8–6 Location of samples in a tank.

A *composite sample* is made of equal portions of two or more spot samples obtained from a tank. The term also applies to a series of line samples obtained from a free-flowing pipeline.

A *composite spot sample* is a blend of spot samples mixed in equal proportions for testing. Tests may also be made on the spot samples before blending and the results averaged. Spot samples from crude-oil tanks are obtained in several ways. A three-way spot sample is taken on tanks larger than 1,000-barrel (160 cubic meters) capacity that contain in excess of 15 feet (5 meters) of oil, although it may also be used on tanks of smaller capacity. Samples should be taken at the upper, middle, and lower or outlet connections, in that order. A two-way spot sample is taken on tanks larger than 1,000-barrel capacity that contain in excess of 10 feet (3 meters) and up to 15 feet of oil, although it may also be used on smaller tanks. Samples should be taken at the upper and lower connections, in that order. A middle spot sample is taken on tanks larger than 1,000-barrel capacity containing 10 feet or less of crude oil. One spot sample should be taken as near the center of the vertical column of oil as possible.

A *tap sampling* may be done through sample taps. The tank should be equipped with at least three sampling taps placed equidistant throughout the tank height and extending at last 3 feet (just less than 1 meter) inside the tank shell. A standard ¼-inch pipe with a suitable valve is satisfactory.

Samples of crude petroleum may also be taken through sample cocks properly placed in the shell of the tank. The upper sample cock should be located 18 inches (457 millimeters) below the top of the tank shell; the lower sample cock should be located level with the bottom of the outlet connection or at the top of an upturned elbow or other similar fitting if installed on the outlet connection. The middle sample cock should be located halfway between the upper and lower sample cocks. An additional cock for the clearance sample should be located 4 inches (approximately 100 mm) below the bottom of the outlet connection to determine whether the level of saleable oil is at least below this point.

The sample cocks should be located a minimum 6-foot (1.8 meters) distance circumferentially from the pipeline outlet and drain connections and 8 feet (2.4 meters) from the filling-line connection. The sample

DAILY GAUGE REPORT
SPESS DRILLING CO.

Date_____19____

LEASE NAME	TANK NUMBER	TIME OF DAY	YESTER- DAY		TODAY		MADE TODAY	B. S. & W.	GA. AFTER B. S. & W. DR'N	
			FT.	IN.	FT.	IN.	INCHES	DR'N	FT.	IN.

PIPE LINE RUNS

TANK NO.	BEFORE		AFTER		LEASE NAME
	FT.	IN.	FT.	IN.	

FIG. 8–7 A run ticket (courtesy Spess Oil Co.)

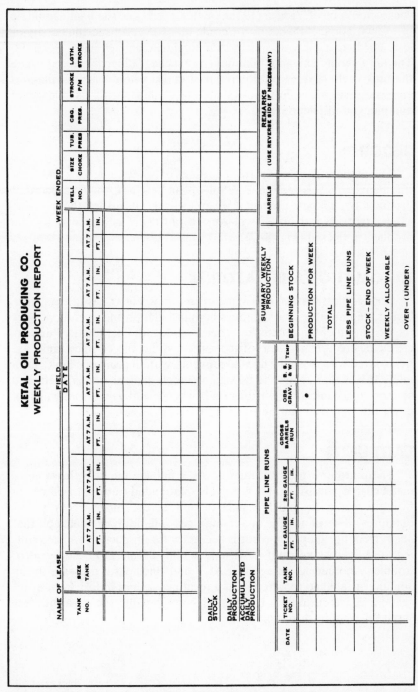

FIG. 8-8 Weekly production report (courtesy Ketal Oil Producing Co.)

cocks should be of ¾-inch (19 millimeters) size, and the lines should be of ¾-inch nominal diameter for crude oil of 18° API specific gravity or less. For lighter oil, cocks of ½-inch nominal diameter should be used. The lines should extend a minimum of 4 inches (102 millimeters) inside the tank shell—except on floating roof tanks, where flush installations are necessary. All sample cocks should be equipped with sealable valves and plugged inspection tees.

RECORDS

The pumper-gauger records the results of his periodic tests and measurements on a *ticket* so the leaseholder can have a permanent record of production and performance. When delivery is made, the pipeline gauger or transport driver will issue a *run ticket* as a receipt for the products delivered (Fig. 8–7).

AUTOMATIC TANK BATTERIES

Today, most leases use a standardized tank battery as to type, size, layout, and fittings. This simplifies both installation and operation. Now the operation of tank batteries is being made more automatic. Pumpers are being freed from some of the more routine chores since tanks are being fitted with overflow lines to prevent overfilling any one tank, automatic valves that divert the flow to an empty tank as one is filled, and electric, mechanical, or air-operated switches and valves that will shut down the well when all tanks are full.

LACT UNITS

There are now fully automatic LACT units in the field (lease automatic custody transfer). They provide unattended transfer of oil or gas from the well to the pipeline since the unit takes samples, records temperatures, determines the quality and net volume, recirculates oil that needs treating, keeps the records necessary for production and accounting, and even shuts itself down and sounds an alarm in the event of trouble. Another advantage of LACT units includes a decrease in the amount of tankage required with a resulting savings in loss from evaporation. Since custody is transferred more rapidly, the lease has less capital tied up in stored oil.

CHAPTER 9

Transportation

There are two phases involved in the transport of petroleum. The first is the moving of the fluids from the well site to the refinery; the second involves the transport of the finished product to the consumer. Petroleum fluids and their byproducts are, in the industry, usually referred to as *product*. This process is really very complex, for the product may actually be transported at or through several refineries, factories, or processing centers before the desired end result is sent to terminals or distribution centers and finally reaches the people who are going to use it.

Petroleum transport may involve pipelines, ocean-going tankers, inland waterway barges, transport trucks, and railway tank cars or a combination of several or all of the means.

THE EARLY DAYS

Drake shipped the product from his well in wooden barrels loaded onto ships or barges or aboard railway flatcars. The birth of the refining portion of the industry increased the demand for petroleum and the railroads, which handled the bulk of the transport at first, soon provided steel tank cars to carry oil from the wellhead to the refinery and the finished product from the refinery to the distributor (Fig. 9–1). As with all other phases of the industry, increasing levels of technology brought about larger, lighter, and vastly improved railroad cars.

One big change in rail freight in general came with the Civil War. Prior to the war, almost every railroad had a different gauge, which meant the cars of one line would not operate on the tracks of another. Thus, goods had to be unloaded and reloaded several times if they were being sent any great distance. In order to facilitate shipment of goods in the North and later when the destroyed southern lines were rebuilt, the present standardization was achieved. This new ability to transfer goods from one part of the country to another led to a rapid growth in rail freight, and up until World War II the railways carried a large percentage of the nations's petroleum.

137

FIG. 9–1 Rail line built into the field to handle crude production around 1914 (Billie Linduff collection).

FIG. 9–2 A modern tanker.

But the use of rail shipment was expensive to the producers and marketers of oil. J. B. Saunders Jr., who was one of the largest of the independent marketers, recalls that he "didn't make any real money until he could switch from rail to barges and save one or two cents per gallon shipping costs." Saunders, who at one time owned 900 tank cars of his own, still had to pay the railways for moving them from one point to another. Other oil men who did not own their own cars had to pay shipping costs plus demurrage (the time a car sits idle at a terminal waiting to be unloaded).

While a considerable amount of petroleum and petroleum-based products are still carried by rail, other forms of transportation have made considerable inroads into rail traffic. The development of the inland waterways system by the Army Corps of Engineers has made relatively low-cost barging available to the industry. Since World War II, pipelines have grown in length, size, and capability. In recent years, the growth of the interstate highway system and modern, large-capacity motor trucks have also carried a large share.

MOTOR TRUCKS

Many small lease-tank batteries are not connected to refinery pipelines, and the producers must depend on tank trucks to pick up their product and deliver it (Fig. 9–2). Trucks have the advantage of being able to go almost anywhere; thus, isolated tank batteries can also be served. Large, modern diesel tractors and semitrailers can carry as much or more as early-day railway tank cars. Trucks also serve the other end of the system as well. Finished products—gasoline, kerosene, jet fuel, and many petrochemicals—leave the refineries in trucks enroute to the customer. Even though product may be shipped from the refiner to the jobber via pipeline, rail, or barge, chances are that it will leave his terminal by truck on its way to the neighborhood filling station.

BARGES

The use of waterways for transportation in the United States dates back to colonial times. The French fur trappers discovered early the value of the Mississippi in getting their goods to market, and cities like St. Louis were founded, grew, and prospered as a result of their proximity to the river. In New York and Virginia, water transportation routes were provided by man-made canals where rivers did not exist or were not suitable. As the English settlers finally crossed the Alleghenies, they found the easiest means of getting their produce to market was via the Susquehanna, Ohio, Kentucky, or Illinois rivers. In the West, the Missouri, Platte, Arkansas, Colorado, Snake, Columbia, and Sacramento rivers became similar avenues of transport.

Since many of the early towns and cities were situated on the rivers, water offered direct point-to-point shipping. In the early days of our nation when roads were largely nonexistent, the rivers and canals were the only means of transporting bulky freight. The canals had the advantage of two-way service. Motive power was supplied by draft animals that walked along a tow path on the bank. Of course, they could pull the canal boats from one town to another and back. However, the canals were expensive to construct and in the north were subject to freezing over in the winter.

The rivers, on the other hand, were free for anyone to use, and many of them remained navigable for much of the year. Their big drawback was that since the vessels used were rafts, keelboats, or canoes that depended on the current for motive power, they were one-way avenues. The coming of the steamboat changed this, though, and for the first time goods and passengers could be carried both ways: from ocean ports such as New Orleans to river towns like Memphis or St. Louis and back.

Today, rivers still provide a great service in transporting bulky, nonperishable goods over great distances at low cost. The steamboat has given way to the diesel towboat that can push a *tow* or string of barges carrying the equivalent of literally hundreds of railway tank cars full of petroleum product. Inland towns such as Tulsa have become port cities due to waterway construction.

Many of our largest rivers have actually been turned into canals by the channelization work of the Army Corps of Engineers. Floods have been controlled by levees and dikes, locks built, and the most modern of navigational aids installed. During World War II, there was great emphasis placed on the inland waterway system. Proponents said that it was much safer to ship the huge amounts of petroleum needed for the war effort inland than by coastal tankers that were vulnerable to enemy submarines. Oil men learned that shipping by barge could be far less expensive than by rail.

This, however, has not been accomplished without controversy. The Corps has come under fire from the railroads and truckers, who see the tax dollars they pay for land and highway use being used to construct a competitive transportation system. Environmentalists claim that channelization has ruined rivers and turned them into sterile industrial canals that still flood periodically but lack the aesthetic benefits of natural streams. Since few members of Congress are likely to vote against multimillion-dollar appropriations for construction in their home districts, it is a dispute in which the correct answers are not clear-cut and one which will not be easily resolved to everyone's satisfaction.

TANKERS

The use of tankers and the size of the ships themselves have increased dramatically in recent years (Fig. 9–3). Carrying oil by sea is certainly not new; the forerunners of today's supertankers were the New England whalers. They would put to sea on voyages that would sometimes last for years in search of the great whales. Once a capture was made, the kill was brought alongside and the blubber stripped off and rendered into oil in pots on deck. The oil was then stored below in wooden casks until the end of the voyage when it was sold for use in lamps.

Whale oil gave way to kerosene, however, and wooden ships to steel, but the need for oil has grown. To help supply the demand, ships have grown increasingly larger. The Allied victory in World War II was won

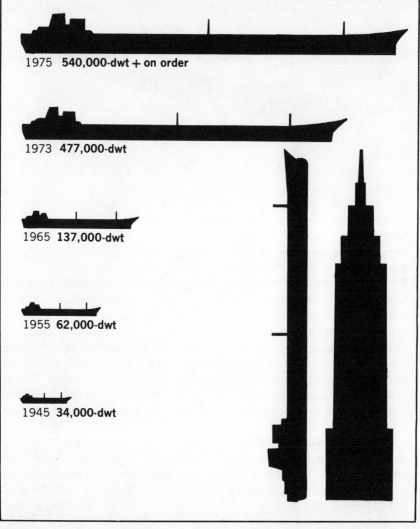

1975 **540,000-dwt + on order**

1973 **477,000-dwt**

1965 **137,000-dwt**

1955 **62,000-dwt**

1945 **34,000-dwt**

FIG. 9–3 How tankers have grown in size (courtesy Mobil <u>World</u>).

due in large part to the fleets of tankers that carried petroleum products from the production areas of the western hemisphere to the combat zones. Those ships, however, with their 34,000 dead-weight tonnage are dwarfed by today's VLCCs (very large crude carriers) of 276,000 DWT or the ULCCs (ultra-large crude carriers) of 540,000 DWT now being built (Fig. 9–4).

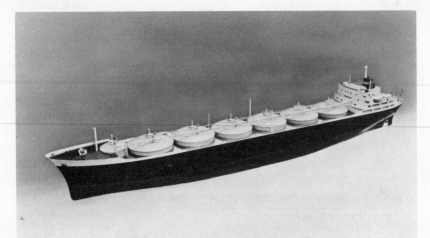

FIG. 9–4 An LNG (liquefied natural gas) carrier.

A 276,000-DWT tanker can carry enough oil in one trip to power a Volkswagen-sized car for 10 round trips to the sun or to heat 22,000 homes for a year. Operators of these huge ships point out that they are safer and more economical than the older and smaller ships.

Since one VLCC can carry many times the cargo of a World War II-sized tanker, fewer ships need to be at sea at a given time, thus reducing the chances of a collision or other accident. The ULCC also carries the very latest in navigational and communication equipment, and its captains undergo special training. The big ships can be built for a lower cost per ton than smaller vessels yet do not require a larger crew to operate them. Their more efficient engines also use less fuel. If the industrialized nations had to depend on the 50,000-DWT tankers used in 1950–1960, an additional 200,000 barrels a day would be needed just to fuel the smaller ships.

In the late 1960s and early 1970s, many operators rushed to construct these superships as the industrialized nations came to depend more and more on imports of oil from faraway producers. However, the increasing price of oil and the Arab oil embargo led to many tankers being taken out of service and a cutback in orders. At the end of 1975, there were 3,674 tankers in the world fleet. During that year, 291 were scrapped, mostly vessels 20 years or more old, and 326 new ones delivered. Projections showed that 610 new tankers would be built by 1980, and that 483 were laid up and out of service. (Layups are a real problem for owners, both in finding suitable berthing space for the ships and also in maintaining them to prevent corrosion and general deterioration while they are not being used.)

Like barges, there has been controversy over the big ships. Environmentalists and commercial fishermen have raised concerns over the impact a rupture and subsequent spilling of hundreds of thousands of barrels of oil could have on marine life and bird populations along the beaches. Both sides have pros and cons, so it remains another of those contemporary questions that will have to be carefully weighed in order to reach the most beneficial and intelligent solution.

The advent of the VLCCs and the ULCCs does not mean the demise of the smaller tankers. They are still needed for short or occasional hauls and for shipments between shallow water ports. A VLCC when fully loaded requires 70 feet of water. Many of the world's ports are not this deep. In fact, there is currently no port in the U.S. capable of handling such a ship. So much of the need of the northeastern U.S. has been served by smaller tankers on the coastal run from the producing areas along the Gulf of Mexico.

One answer to the lack of port facilities is to anchor the big ships offshore and lighten the load onshore by transferring it to barges. This is an unwieldy and time-consuming process that can tie a ship up for as much as two weeks.

Other solutions to the problem would either be to construct deepwater port facilities or to use single-point mooring systems (SPM)(Fig. 9–5). The SPM is a buoy attached to the bottom of the sea and connected to the shore by undersea pipeline. They are located in deep water

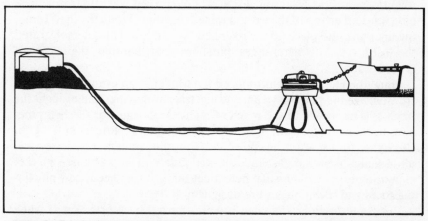

FIG. 9–5 A single-point mooring system (courtesy Mobil World).

away from shoals and other ship traffic and may even be out of sight of land. The tanker attaches a flexible hose to the SPM and is free to swing with the current as her cargo is pumped into shore tanks. In the event a sudden storm should blow up, the ship can quickly disconnect and move away.

As noted earlier, the price paid for oil is largely determined by the API gravity. On such large shipments, this can become an important factor, affecting not only the purchaser and seller but also the vessel itself. There are approximately seven 42-gallon barrels of crude oil per ton. This figure will vary, however, with the API gravity. Both parties then need to know just how much and of what quality the oil is that is being transported. The captain of the ship also needs to know the weight of his cargo (as does the driver of a tank truck if he does not wish to be ticketed for an overweight load). Thus, the oil is tested at the port of embarkation and usually again when it is offloaded. Such testing is usually done by an independent third party, such as the licensed and bonded representatives of the Saybolt Organization of England, who specialize in this type of work.

PIPELINES

As the demand for oil grew, so did the demand for pipelines, which in the coming decades would begin to span the continent. In the early days the lines were laid largely by man and mule power. In 1918 when Harry Sinclair built a 673-mile line from Kansas City to East Chicago, crews of Irish laborers connected the joints of 8-inch pipe by hand using 50-pound tongs as wrenches while the foreman tapped out a rhythm on the pipe with a sledge. Racing eastward to beat the weather, the crews could lay 206 joints in a 9-hour shift. Their objective was to cross the Mississippi at Fort Madison, Iowa, while the river was still frozen. The pipe was laid on top of the ice and joined together. When the thaw came, the line sank into place on the river bed. World War I also demonstrated the usefulness of natural-gas pipelines and boosted their growth (Fig. 9–6).

In 1908, Henry Ford introduced the Model T, an event that signaled the doom of the popular steam and electric automobiles and brought the price of a car down to the reach of many Americans for the first time. Two years later, there were nearly half a million motorcars in the U.S. But by 1920, just after the end of the War, nine million were in service, all of which required gasoline, oil, and grease. This was a new market clamoring for petroleum. To help meet this growing need, new pipelining tools and techniques were designed.

These technological advances were spurred by World War II when coastal tankers were being sunk by German submarines in sight of the

FIG. 9–6 Pipeline in the 1920s (Billie Linduff collection).

crowds along the East Coast beaches. In order to safeguard the vital supply of war fuel, the government authorized the construction of the 22-inch Big Inch and its companion 20-inch Little Inch pipelines from Texas to the East Coast. Both lines marked a signal feat in pipeline construction; never before had lines this large been laid such a distance so rapidly.

Today, pipelines fall into two general categories: oil, which may be used for other products besides crude, and gas, which are used for the transmission of natural gas. There are several types of each used for different purposes. Construction methods differ between the two. Each is responsible to a different regulatory agency, and even the method of land acquisition is different. Both have the same goal, however: to transport petroleum liquids as efficiently and cheaply as possible.

OIL PIPELINES

Oil pipelines may be used to transport crude oil or finished products such as gasoline of various grades, fuel oil for home or industrial heating, and many other derivatives and commodities (Fig. 9–7). Modern lines are built to deliver a multitude of products in rapid sequence. Separation of products is carried out by several methods. Each shipment is called a *batch*. In years past, the batches may have been separated from each other by running water or kerosene between them.

Today compatible crudes or products may well be sent one behind the other in special batching sequences. The points where one stops and

FIG. 9-7 A modern pipelining crew.

another begins is known as a *batch interface*. These can be detected by *recording gravitometers* or by sampling. Batch interfaces can also be detected by radioisotopes or dyes. The isotopes offer the advantage of electronic detection; therefore, each batch can be tracked all through the line. An inflatable *rubber sphere* may also be sent through the line to separate different products.

Pressure created by pumps at pumping stations placed at intervals along the line provides the power to move the product on its way. In former years, pipeline operators owned their own telegraph companies, which provided communications along the line. Personnel in each pumping station controlled the flow of product by means of manually operated valves. Today telegraph lines have given way to microwaves, and the flow of product is highly automated and computer controlled.

Stations may be from 80 to 150 miles apart. The spacing depends on the diameter of the line, the terrain, and the type of product to be moved. The flow rate and pumping pressure are closely monitored and controlled, and automatic and manual warning and alarm systems are installed to signal either drops below or rises above the predetermined levels. These systems can shut the line down if an abnormal situation develops. Since many stations are now totally automated, an operator at a central location can control all of the pumps, valves, compressors, and regulators for an entire pipeline system and monitor each shipment.

Pipelines, like any mechanical device, must be carefully maintained. The first step in proper maintenance is the cleaning of the line, which

FIG. 9–8 A pipeline pig or scraper.

must be done at regular intervals. A device called a *pig* or *go-devil* is pushed through the line ahead of a shipment to scrape away any deposits of dirt or wax that may have accumulated on the inside walls. It can be removed at a pumping station, cleaned, and replaced in the line to continue its journey (Fig. 9–8).

Interstate oil pipelines are common carriers; as such, they come under the regulations of the Department of Transportation. Among other things, this means that the owners of the line must publish their tariffs (rates and regulations) and, if they ship their own product through the line, must pay the same price as they charge others. They must accept oil for shipment from any company as long as the product meets the conditions for all customers published in the tariffs.

Oil pipeline builders must negotiate individually with the owners of the land they will cross. A one-time fee is paid for the use of the land for pipeline right-of-way. This can range from as low as $2.50 to as high as $40.00 per rod (16½ ft). As with a drilling contract, an agreement is signed by both parties, spelling out their rights and responsibilities. On the part of the pipeline company, this may include provisions to restore the land after construction, pay for crops lost due to construction, spare specified trees or structures, bury the line below a minimum depth (usually three feet), and replace fences, gates, and cattle guards.

Arrangements also have to be made for crossing points with the owners of other pipelines that may be encountered along the route. The new line always crosses under the old. At intersections such as at town limits, as many as four or more lines may cross, with each newer one passing under the next oldest above. The property owner, for his part, guarantees the pipeline company rights of transgress and egress to his property.

Oil pipelines fall into three major categories: gathering, transporting, and distributing.

Gathering Lines

The pipelines used to transport crude from individual production units or leases to a central point are called the *gathering system* (Fig. 9–9). These may feed to a refinery or port or into a larger trunk line that carries the production of a number of gathering systems to a refinery. Normally, gathering pipelines are from four to twelve inches in diameter and may be anywhere from a few feet to several miles long. Although the lines are relatively small and short, their number and complexity makes them a major portion of the pipeline industry. Testing must be done on the quality and quantity of the oil coming in from each lease for proper custody transfer and so each owner will get fair payment for his oil.

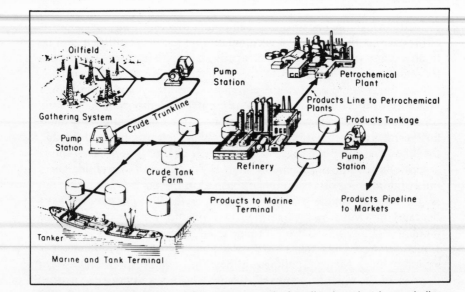

FIG. 9–9 A pipeline system composed of gathering, trunk, and distribution lines (courtesy Petroleum Extension Service).

Trunk Lines

The gathering systems feed into larger trunk lines that may carry the oil from any number of gathering systems to central storage, port, or refineries. Several parties may share in the ownership of a trunk line, and many of these lines are also highly automated. Since many producing fields are located in remote areas, this network of gathering and trunk lines that offer efficient and low-cost transportation from well to refinery are a mainstay of the petroleum industry.

Distribution Lines

Distribution systems, or product lines, are those that run from refinery to user. Their engineering must be more concise due to the varied number of products they carry in contrast to the gathering and trunk lines. Precise separation of batches is necessary to maintain the integrity of the individual products being shipped. Therefore, while this is the youngest member of the pipeline industry, its technology has had to develop much more rapidly.

Distribution lines are the reverse of the others since they begin as larger lines and decrease in size as branches feed off them to serve product users. They usually run into areas of high population concentrations, so public safety is a prime consideration in their construction and operation.

GAS PIPELINES

While technically speaking a piece of pipe can carry either a gas or a liquid, there are certain pertinent differences between the two. First, a gas line built to transport natural gas in interstate commerce is considered a public utility, not a common carrier. This places it under the regulation of the Federal Energy Regulatory Commission (FERC), which sets the rates that the owners of the gas can charge for their product in states other than those in which it was produced. This has been another area of controversy in recent years, with consumers wanting regulations concerning rates maintained and producers wanting them removed. Again, this represents a problem not easily solved. The users, of course, want to purchase gas as cheaply as possible. The producers say that low rates do not allow them to carry out exploration, development, and production of new sources of natural gas.

As public utilities, gas pipeline owners enjoy the right of eminent domain when it comes to acquiring right of way for their lines. This means that the company can purchase the ROW, but at the going market price. If the company and the property owner cannot agree on a figure, then it is set by a court.

Instead of pumps and pumping stations, compressors and compressor stations are used to move the gas through the lines. Construction techniques are also different. For example, pipe diameters may have to be changed at different points on an oil line to allow for the weight of the hydrostatic head. At the bottom of a mountain grade or river crossing, increased-diameter pipe must be used on an oil line to prevent rupture from the weight. This is not necessary with a gas line.

CONSTRUCTION

While all pipelining techniques have some common characteristics, the environment will dictate the final cost, the complexity and technology, and even the choice of contractors to build the line. Lines are laid on land, in marshy or swampy areas, and far beneath the sea. Each calls for contractors and machines that specialize in that particular area.

As with the drilling of a well, the construction of a new pipeline begins with careful, detailed planning of both the market and the reservoir. Are there both a supply of and demand for the oil that will economically justify the cost of building a pipeline? Next come the engineering studies. Potential routes are surveyed by aerial photography and surface mapping. Terrain features such as rivers, swamps, mountains, and other obstacles are taken into account, as are the availability of access roads, pipe yards, and supply depot sites. Engineers skilled in fluid mechanics and hydraulics determine the size of the pipe based on viscosity of the product to be transported, gradients in the terrain, flow, pressure, and other variables.

Only when all of these many factors have been taken into consideration plus an environmental impact study made and approved by the Environmental Protection Agency and any historical site or archaeological studies that may also be required completed can actual steps be taken toward construction.

The Spread

All of the men and machines used in the construction of a pipeline are referred to as the *spread* (Fig. 9–10). The actual number of personnel and types and amounts of equipment will be dictated by the job at hand. On some jobs, the spread will handle all phases of construction. On others, specialized contractors may be called in for various operations, such as right-of-way clearing, heavy hauling, river or highway crossings, coating and wrapping, and testing. In charge of the day-to-day operations will be the spread superintendent, an assistant, office manager, timekeeper, payroll clerk, engineer, and supplyman.

FIG. 9–10 A pipeline spread.

TABLE 9–1 Personnel spread for a large pipeline. (Petroleum Extension Service)

Position	No.	Total
Field Office and Overhead		
Spreadman	1	
Assistant spreadman	1	
Office manager	1	
Timekeeper	1	
Purchasing agent	1	
Material man	1	
Heavy-duty mechanics	1	
Truck mechanic	4	
Utility welder	1	
Fuel truck driver	1	
Fuel truck helper	1	
Grease truck driver	1	
Grease truck helper	1	
Utility truck driver	1	
Lowboy truck driver	1	
Float truck driver	1	
Night watchman	1	20
Right-of-way		
Clearing and fencing foreman	1	
Straw boss	1	
Dozer operators	2	
Truck drivers	2	
Power saw operators	6	
Laborers	25	37
Grading		
Foreman	1	
Dozer operators	3	
Laborers	3	7
Ditching (Dirt)		
Foreman	1	
Straw boss	1	
Ditching machine operators	2	
Oilers	2	
Stake setters	2	
Dozer operator, ripper	1	
Laborer	1	
Backhoe operators	4	
Oilers	4	
Clam operator	1	
Oiler	1	20
Ditching (Rock)		
Blasting		
Sideboom operators	2	
Wagon drill operators	4	
Dozer operator	1	
Powdermen	2	
Truck drivers	2	
Jackhammer men	2	
Laborers	6	19
Road Crossing		
Straw boss	1	
Road boring machine operator	1	
Sideboom operator	1	
Truck driver	1	
Dope pot fireman	1	
Laborers	5	10
Stringing		
Foreman	1	
Crane operator	1	
Truck drivers	7	
Sideboom operator	1	
Tow cat operator	1	
Laborers	4	
Mechanic	1	16
Bending		
Foreman	1	
Engineer	1	
Rodman	1	
Helper to engineer	1	
Sideboom operator	1	
Bending machine operator	1	
Swampers	2	
Laborers	2	10

TABLE 9–1 Personnel spread for a large pipeline. (Petroleum Extension Service) (continued)

Position	No.	Total
Pipe Gang (Alignment and Stringer Bead)		
Foreman	1	
Straw boss	1	
Sideboom operator	1	
Swampers	4	
Stringer bead welders	3	
Hot pass welders	4	
Welder helpers	9	
Tack rig operators	2	
Stabber	1	
Spacers	2	
Clamp man	1	
Truck drivers	2	
Laborers	10	42
Welding Gang		
Foreman	1	
Welding steward	1	

Position	No.	Total
Welders	10	
Welders helpers	12	
Tow tractor operator	1	
Bus driver	1	
Graded apprentice	1	27
Cleaning, Priming and Coating Gang		
Foreman	1	
Cleaning machine oper.	1	
Coating machine oper.	1	
Sideboom operators	3	
Straw boss	1	
Pot fireman	5	
Paper latchers	2	
Tow tractor operator	1	

Position	No.	Total
Truck drivers	2	
Laborers	17	34
Lower In		
Foreman	1	
Sideboom operators	4	
Clam operator	1	
Oiler	1	
Dozer operator	1	
Truck drivers	2	
Laborers	10	
Pot fireman	1	21
Tie In		
Foreman	1	
Sideboom operators	1	
Welders	2	
Welder helpers	2	
Truck driver	2	
Laborers	6	
Clam operator	1	

Position	No.	Total
Oiler	1	
Pot fireman	1	17
Backfilling and Cleanup		
Foreman	1	
Straw boss	1	
Backfiller operator	1	
Dozer operators	3	
High lift operator	1	
Farm tractor operator	1	
Truck drivers	2	
Laborers	15	25
GRAND TOTAL		305

TABLE 9–2 Equipment spread for a large pipeline. (Petroleum Extension Service)

Field Office and Overhead	
Cars	2
Pickups	7
Welding rig	1
Fuel truck	1
Grease truck	1
Utility truck	1
Tractor w/lowboy	1
Tractor w/float	1

Right-of-way, Clear, Burn, and Fence	
Pickups	2
Dozers	2
Winch truck	1
Flatbed truck	1
Power saws	6

Right-of-way, Grade	
Pickup	1
Dozers	3

Ditch (Dirt)	
Pickups	7
Ditching machines	2
Dozer w/ripper	1
Clamshell	1

Ditch (Rock)	
Sidebooms	2
Wagon drills	2
Compressors (900)	2
Trucks	2
Jackhammers	2

Road Crossings	
Pickup	1
Boring machine	1
Sideboom	1
Winch truck	1
10 bbl. tar kettle	1

Stringing	
Pickup	1
Crane hoist	1
Tractors w/pipe trailers	7
Sideboom	1
Tow tractor	1

Bending	
Pickups	2
Sideboom	1
Bending machine	1
Transit & level rod	1

Pipe Gang	
Pickup	2
Tractor-welders	2
Sidebooms	2
Trucks	2
Gang bus	2
Set inside clamps	1
Buffing rig	1

Welding Gang	
Pickup	1
200-amp welders	11
Tow tractor	1
Gang bus	1

Dope Gang	
Pickups	1
Sidebooms	4
Tow tractor	3
Clean & Prime machine	1
Coat & Wrap machine	1
27 bbl. tar kettles	1
Winch truck	2
Flatbed truck	1
Sets, cradles & strongbacks	1
Holiday detectors	1

Lower In	
Pickups	2
Sidebooms	4
Clamshell	1
Dozer	1
Trucks	2
27 bbl. tar kettle	1
6" water pump	1

Tie In	
Pickups	2
Sidebooms	2
300-amp welders	2
Truck	1
Clamshell	1
10 bbl. tar kettle	1
6" water pump	1

Backfill and Clean Up	
Pickups	4
Backfiller	1
Dozers	3
High lift	1
Farm tractor	1
Dragline	1
Winch truck	2
Flatbed truck	2

Large spreads may require as many as 500 workers and cost $2 million or more to equip. On mammoth jobs such as the Alaska pipeline, which cost more than $7 billion to complete, the list of men, materials and machines needed was almost endless. Mountains call for extra bulldozers to steady the clamshells (dredgers) while trenching and filling; forests require chain saws; and rocky terrain demands rippers, wagon drills, and explosives. So there is no such thing as an average spread.

At the beginning, the spread superintendent has three documents to guide him. Maps show in detail every terrain feature that might affect construction. The line list shows all the property to be crossed, the owner, and any special considerations as well as permits for river, highway, and railway crossings. The contract specifications spell out just what is to be built, how, and what materials are to be used.

RIGHT OF WAY

The first step in cross-country construction is the clearing and preparation of the right of way. The clearing crew must open fences and build cattle guards, gates, and bridges as needed. A strip from 50 to 75 feet in width is cleared and graded. This may call for the removal of timber, which is salvaged for sale. The removal of underbrush, stumps, and large rocks that would interfere with grading is also necessary.

After the right-of-way is cleared and leveled, the ditch for burying the pipe is dug. It must be deep enough to meet the minimum depth requirements and also at least 12 inches wider than the pipe. The path has already been surveyed and marked out with a row of stakes, so under ideal conditions the trenching machines can move ahead rapidly, throwing the spoil just to the left of the trench where it can be used for fill. In wet ground, clamshells will have to be used, and in rocky terrain rippers pulled behind bulldozers, wagon drills, and explosives may be called for.

Stringing. The pipe is furnished by the owner of the line and shipped to locations along the path of the pipeline. From these stockpiles the owner may deliver it or the contractor may have someone haul it to the line.

It may be decided to yard coat the pipe before final delivery. All of the pipe must be cleaned, wrapped, and coated before burial. This may either be done onsite or at a pipe yard before the pipe is brought to the ditch. If yard coating is decided upon, several yards may be set up along the route. Here the pipe will be cleaned, primed, coated, wrapped, and double-jointed (two sections welded together), with only enough of each end left exposed to permit welding (Fig. 9–11).

Crossing. Roads, railroads and other rights of way encountered are usually crossed by boring under them. This way, traffic is not disrupted

FIG. 9–11 Coating and wrapping before the pipe is laid in the trench (courtesy Petroleum Extension Service).

nor does the surface have to be patched. Boring machines use an auger bit to drill a horizontal hole. As the hole is being drilled, a casing is put in place through which the pipe is run. The casing prevents cave-ins, settling, and damage to the road bed above (Fig. 9–12).

Bending. Every time the ditch changes direction or elevation, the pipe must be formed to fit it. Thus, many joints of pipe must be bent or curved. Small pipe can be formed around a bending shoe, but large-diameter and thin-wall pipe must require special handling if it is to be

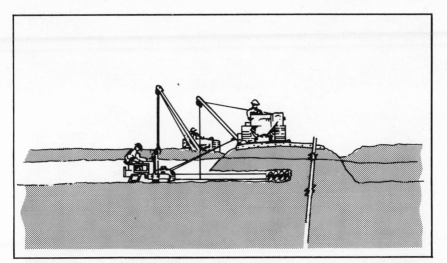

FIG. 9–12 Boring under a highway crossing and pushing the casing in place.

kept round and distortion free. Special mandrels and bending machines have been developed for this purpose.

Pipe Gang

The pipe gang is the largest crew of the spread. They begin their work when the pipe is laid alongside the ditch. They must make sure that the pipe ends are clean and ready for welding. All scale, rust, and coating material must be cleaned off with brushes and buffers. If the end has been damaged in shipping, a new bevel must be cut so a good weld may be made.

The joints are welded together one at a time. For pipes larger than 12 inches, internal-alignment clamps must be used. For smaller pipe, external clamps squeeze them into roundness and alignment for welding.

During this process, the pipe is positioned by sideboom tractors while the first or *hot pass* is made. Then if the diameter of the pipe requires it, another pass is made. In between each, the bead must again be brushed and cleaned. On good terrain and in good weather the pipe gang may complete 250 or more initial welds per day.

Welding Gang

After the pipe gang has made the initial hot passes, they will be followed up by the welding gang, which will make the final welds. The number of welds made by the *firing line,* as the welding gang is known, will depend on the method of welding and how thick the walls of the pipe are (Fig. 9–13).

FIG. 9–13 An automatic welding machine.

When the two beveled ends of the pipe joints are brought together, they form a V-grooved channel. The desired end result is a smooth joint, so enough beads are deposited to achieve this. A completed weld then may take from two to five passes: stringer bead, hot pass, filler bead, and cap bead. Any irregularities must be removed before the next pass is begun.

Either stick or wire welding may be used. *Sticks* are the common electrodes or welding rods used in arc welding. Wire welding uses a wire-type electrode supplied to the welding gun on a spool. Wire welding may be enhanced by the use of CO_2 during the process.

Inspection

Proper inspection techniques, including X-raying the welds, helps prevent leaks later on and cuts down on the high cost of digging the pipe up for repair (Fig. 9–14). Some flows can be detected by visual inspection of the welds; others can be spotted only by X-ray. The nature of the line itself will determine how many X-rays are actually needed. If it is a high-pressure or thin-wall line, all of the welds may be X-rayed as are those that will be used under water or in railroad or highway crossings. Other situations may only require that 25 percent be so inspected.

FIG. 9–14 Examining X-rays made of welds. Both internal and external X-rays are made (courtesy Petroleum Extension Service).

Portable X-ray machines are used on both the interior and exterior of the pipe. The films are processed and read in the field.

Dope Gang

After welding, the pipe is ready for cleaning, priming, coating, and wrapping. If the pipe has been previously prepared in a pipe yard, only the welds need to be so treated. If not, then the entire line must be prepared for burial.

Cleaning and priming machines scrub the pipe to remove all foreign matter and apply a coat of enamel primer. Cleanliness is vital, because dust or dirt remaining on the pipe under the primer will leave a *holiday* or flaw in the protective coating, which will have to be repaired later (Fig. 9–15).

When the primer has dried, a second machine moves along that cleans the dust from the paint, applies a coat of tar, and imbeds a covering of glass fiber. Hot coating material is sprayed over this and finally

FIG. 9-15 Repairing a holiday.

the pipe is wrapped with heavy kraft paper and sometimes a plastic wrap.

An electronic holiday detector, sometimes called a *jeep,* is then passed over the pipe to locate any hidden flaws. The coating may be allowed to cool and harden before the pipe is buried.

Lowering

The pipe may be lowered into the trench as part of the coating and wrapping process. *Cradling,* as this is called, reduces possible damage to the coating by eliminating an extra step in handling. Or the coating may be allowed to harden before the sidebooms pick the pipe up from the wooden skids and lower it into the trench. This is usually done late in the evening when the metal is contracted.

Tie-ins. There may be numerous spaces left in the line at crossing points and rivers between the line itself and the sections laid under the river, railroad, or highway. These are joined by the tie-in crew, which is really a miniature spread itself. They perform all the wrapping, coating, and welding chores of the regular crew.

River crossing. Sections of the line that cross rivers are usually built by specialized contractors. They may either be sunk in the river bottom or carried across above the surface much like a suspension bridge. Pipe may be joined onshore and pulled across, or it may be lowered from barges. Weights are added as needed to prevent the line from floating.

Controls. When all of the above steps have been completed, provisions are made to set in valves, bypass lines, and scraper traps. These last run parallel to the line and are used for insertion of pigs, scrapers, and batch separators.

Filling. Part of the trench was filled as the pipe was lowered to help hold it in place. Now dozers move in to cover the rest of the trench and efforts are made to place the topsoil back on last so that vegetation will soon take hold and grow again. The right of way is restored as closely as possible to its original state. Fences and contours are restored, trash is removed, and the right of way made ready for planting if desired.

CORROSION CONTROL

Corrosion is a process in which metals tend to change, under certain conditions, into other forms such as oxides (rust). Most metals and alloys can and usually do corrode. The tendency is stronger in some metals than others.

Iron in the form of steel is the major metal of which our underground and underwater pipelines are made of. Metallic iron is produced from ores or iron oxides. Its metallic state is not stable, and natural forces tend to return it to its more stable oxide form. In other words, if iron is exposed to air and water, oxidation or corrosion occurs.

External corrosion of metallic pipelines is an electrochemical phenomenon, and oxygen in some form is necessary. Often terms such as *electrolysis, oxidation,* and *chemical attack* are used. These are not different processes; they are different aspects of corrosion.

Before external corrosion of pipelines can occur, there are certain conditions that must be met.

1. The pipeline must be in contact with an electrolyte containing ionized molecules, both positively charged hydrogen ions and negatively charged hydroxyl ions. This could be moist soil or water surrounding the pipeline.

2. One or more portions of the wet pipeline surface must be anodic and other portion or portions cathodic. The *anodic portion* of the pipeline means that it is an anode (the electrode at which oxidation or corrosion occurs). The *cathodic portion* is the cathode (the electrode, opposite of anode, a gain of electrons).

3. An electrical potential between the anode and cathode must exist. These electrical potentials are normally very small.

4. There must be a solid conductor connecting the anode and cathode. The metal pipe would be the conductor.

5. An electrolytic path must exist between the anode and cathode. This could be the moist soil or water surrounding the pipeline.

Generally corrosion is more extensive on the bottom of the pipe where more moisture is likely, especially with bare underground pipelines. Uncoated and poorly coated pipelines are more likely to corrode than well-coated pipelines.

Dissimilar surface condition of the pipe can cause corrosion. The metals are not pure and will scale, or mechanical scratches will cause dissimilar surface conditions. Action of corrosion cells themselves cause changes in metal surface conditions.

PROTECTIVE COATINGS

As stated earlier, the pipeline must be in contact with an electrolyte containing ionized molecules to corrode. If it were possible to coat the pipeline with a material that was absolutely waterproof and free from holes (holidays), corrosion would be stopped. At this time, there is no perfect coating that has these two properties that will remain on the pipe permanently. An approach to perfect coating is very expensive, so it is obvious that imperfect coatings must be used. Experience has shown that a combination of a reasonably good coating plus cathodic protection will retard corrosion. A good coating should have many special properties, including a high electrical resistance, resistance to water, and an ability to withstand mechanical damage.

There are many coatings used on pipelines, each having its merits and restrictions. Some of the more common coatings include vinyls, epoxies, chlorinated rubbers, alkyds, phenolics, asphalts, and coal tars. The method of applying coatings will vary depending on the situation. Some are hand-applied, while others are machine-applied either in the yard or over the ditch.

Coating thicknesses vary widely, each depending on the material used and method of application. A typical application including the primer and hot enamel is about 0.10 to 0.15 inches thick plus the wrapping thickness. Thermoplastic resins and tapes are in the 0.010 to 0.030 inch range.

It is very important that the coating be free from holidays because a small hole in the coating could concentrate corrosion there. The amount of metal corroded at the anode is approximately 20 pounds per ampere per year. This means that 1 ampere of direct current discharged for 1 year from iron into an electrolyte would dissolve 20 pounds of iron. It is obvious that a very small amount of direct current could very quickly cause a leak in a pipeline that has holidays in its coating with no means to retard or control it.

Protective coatings are expensive and can range from 5 to 10 percent of the total pipeline construction cost. But compared with replacing part or all of the pipeline, it is cheap.

CATHODIC PROTECTION

Cathodic protection is an effective means of retarding corrosion on underground or underwater pipelines. There are basically two types of systems: the galvanic anode type and the rectifier type. Each type has characteristics that makes it more adaptable to a given problem than the other. Each is effective both on bare or coated underground or under-water pipelines. Both are more efficient on well-coated pipelines. The two types make all parts of the pipeline become a cathode, thus the term cathodic protection.

Galvanic anodes require no outside source of electrical power for operation. As the metals in the anodes corrode or oxidize back to their more stable forms, electric power is generated. For this reason, they are often referred to as sacrificial anodes. They are molded pieces of magnesium, zinc, or aluminum metal and are buried in the soil with insulated solid conductors connected to the pipeline. Due to the differences between the metal in the anodes and iron in the pipe, a small voltage exists. When placed in an electrolyte such as moist soil, voltage will flow from the metal with the higher potential to metal with lower electrical potential. The anode will corrode to protect the iron cathode. However, their area of influence is relatively small and many are required to protect a long pipeline with a large diameter.

The rectifier type of cathodic protection requires an outside source of electrical power for operation. A typical system takes alternating current (AC) from a conventional power line, passes it through a rectifier to change it into direct current, then feeds it into anodes buried along the pipeline (Fig. 9–16). Solar panels are used effectively at very remote sites to provide the necessary power source.

Rectifier types are more often used on larger pipelines because of the larger area of influence they cover. It is not uncommon to protect 20 miles of pipeline with each rectifier. They do require external power that is expensive.

Rectifier types normally use graphite or high silicon cast-iron anodes that are buried to form a ground bed. When graphite anodes are used, they are surrounded by a coke breeze backfill. The number of anodes

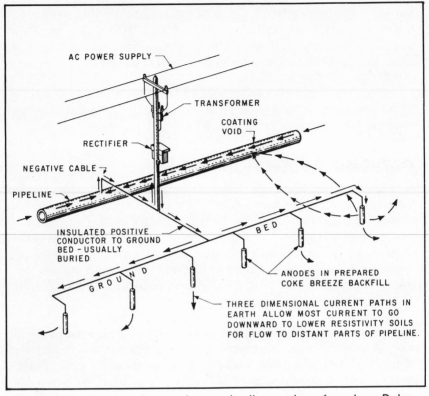

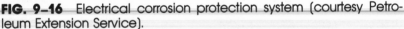

FIG. 9–16 Electrical corrosion protection system (courtesy Petroleum Extension Service).

used will vary with the situation; normally 20 anodes are used. The calculated life of the ground bed is usually 20 years.

Cathodic protection and coatings retard external corrosion of pipelines but do not protect the internal surface. Internal corrosion is not a problem for many crude oil pipelines. Where internal corrosion is a problem in oil or gas pipelines, they are usually protected by coatings. The inside of the pipe is sandblasted and cleaned then sprayed with a thin film of coating. The coating is the only protection used internally because there is no practical way to install anodes for cathodic protection.

PIPELINE IN SWAMPS AND OFFSHORE

Construction in swamplands is like on land except barges are used. Dredges open channels. Flotation devices may have to be provided for the line until it is ready to be lowered from the lay barge, and then weighting attached to hold it in place on the bottom. As one section of

FIG. 9-17 A pipe lay barge.

pipe is completed, it is lined up with a new one on the deck of the barge
and then pushed into place. Construction in such an environment combines elements of both land and offshore construction and is handled by
specialized contractors.

Undersea pipelining is another of those areas of the petroleum industry that has seen rapid technological growth in the past few years. New-
generation lay barges and ships are now able to perform feats thought
impossible in the recent past. The development of the North Sea oil and
gas fields has brought much in the way of new tools and techniques (Fig.
9-17).

Ships and barges are now capable of laying more than five miles of
pipe per day in more than 600 feet of water and in weather that might
produce 15-foot waves. Again, many of the procedures are the same as
on land, but all operations must be carried out aboard ship. Hundreds of
miles of lines have thus been constructed to bring the much-needed
petroleum ashore. Shifts of divers may work around the clock, and jet
sleds are used to blast out trenches in the bottom.

One major problem encountered, particularly along the rocky shores
of the North Sea, is getting the pipe through the surf line where the water

is too deep for land equipment and too shallow for the lay barges or ships. This has been solved to some extent by completing the surf section on land and then pulling it out to sea to be joined. Pipelines now lie under all the world's seas.

PIPELINE MANTENANCE

Once laid, the pipeline must be inspected on a regular basis. In years past, much of this was done by the line walker who would walk the length of the line looking for leaks, irregularities, and even new construction projects along the right of way that might affect the line. Today much of the visual inspection work is carried out from light aircraft or helicopters that fly the route at low altitudes. In areas that are inaccessible to low-flying aircraft, inspection by foot or land vehicle must still be carried out.

THE ALASKA PIPELINE

The Alaska Pipeline, finally placed in service in 1977 at a cost of more than $7 billion, is such a noteworthy engineering feat that it deserves special mention. Never before had a pipeline been built so far in such a harsh climate or over such difficult terrain. Not only did all of the past experience in pipeline construction have to be called on, literally hundreds of new techniques had to be devised.

Running more than 800 miles south from the North Slope at Prudhoe Bay, it is designed to deliver 1.5 million b/d (barrels per day) to the tanker terminal at Valdez. It crosses three mountain ranges, 250 rivers and streams, and almost 400 miles of permafrost and tundra.

Due to the climate, the balance of nature in the tundra region hangs very delicately. Any damage is almost irreparable. At the same time, a temperature change of as little as one degree can cause the permafrost to melt. When it refreezes, the pressure causes the earth's surface to buckle upward, destroying any man-made structure placed directly on the permafrost.

Thus, much of the pipeline could not be buried but was suspended above the surface on supporting legs. These legs are filled with ammonia, which cycles up and down, keeping the temperature constant. Provisions have also been made to control the temperature of the oil—not only to keep it from doing environmental damage from too high a temperature while it is moving, but also to keep it from freezing and breaking the line if there is a stoppage of flow. Almost perpendicular mountain slopes had to be transversed, and an access highway had to be built to get the men and equipment into the job sites. All of this had to be done when temperatures dropped as low as 60 degrees below zero. In addition to the technological problems to be overcome, progress was often halted by the lambing season of Dall sheep, the spawning season of salmon, and other ecological concerns.

The pipeline makes available the estimated 9.6 billion barrel reserves of the North Slope. At almost the same time that it was placed in service, announcement was made of plans for another, this one to carry natural gas from the Prudhoe Bay area over a much longer route, angling southeast across Canada down toward the middle of the United States.

ADVANTAGES OF EACH METHOD

Each mode of transportation has its own peculiar advantages. Pipelines, despite their tremendous construction costs and complexities, now span hundreds of thousands of miles and carry millions of barrels of petroleum fluids daily at a cost far lower than any other means of transportation. This is vital when a difference in price of a few cents per gallon can mean the difference between profit and loss. From their humble beginnings at Oil Creek, pipelines have grown to the point that they transport almost half of all petroleum products.

FIG. 9–18 A tanker used on railroads.

More efficient towboats and our vast system of inland waterways make possible tows that can transport as much oil in one trip as 30 or 40 railroad trains. In areas accessible to the rivers, they provide a most economical means of petroleum transport.

Great improvements have been made in rail transport also. Newly designed tank cars have shed as much as 20,000 pounds of excess weight, and their fiberglass and alloy tanks can carry more than ever before. Rail lines are available to most refineries and serve population centers where much of the product is consumed (Fig. 9–18).

Tank trucks are now capable of carrying 13,000 gallons in a single load and, while more expensive to operate, have the ability to go almost anywhere. Thus, at one end of the process they can pick up crude from isolated leases and at the other end deliver finished product to the most isolated consumer.

CHAPTER 10

Refining

The practice of refining is even older than the petroleum industry itself. This began with the simple distillation of many different raw materials in Europe in the 18th century to see if new products could be obtained. Coal, shale, tar from seepages, and even whale oil were experimented with.

By 1859 there were some 80 coal-oil plants in the United States, but many of these quickly shut down after the advent of the Drake well and rapid developments in the infant industry.

The first new refinery after Drake's discovery was built near Oil Creek in 1860 by William Barnsdall and William H. Abbott. They were followed by many more. By the end of the Civil War, more than 100 plants were using 6,000 barrels of crude per day. By 1870 a basic operating pattern had developed.

In a typical batch run, the crude would begin to vaporize at 180°F. The temperature of the still would gradually be brought up to 1,000°F. The lightest product or fraction (the first portion of the crude to vaporize) was gasoline of 72–74° API gravity. The next was a 62–65° API naphtha or benzene. There was almost no market for these products, so they were either burned or dumped as waste.

The next *cut* or fraction was 40–50° API gravity kerosene, which was the principal product of the day. The residue left after the gasoline, naphtha, and kerosene cuts were made was treated with acid and naphtha and run again with steam-refined stock. This was further treated with a lubricating material known as bright stock. Other stocks were obtained from 32–36° API gravity distillates. Petroleum greases were made with fatty oil and wax.

The early kettles soon gave way to *shell stills* (Fig. 10–1). These consisted of a battery of vessels through which the crude or feedstock passed. Each vessel had a higher temperature than the previous one. The distilling process broke the crude down into different cuts. The *lighter ends* or *aromatics* were those that vaporized first at lower temperatures. The shell stills in some areas were replaced by large cylindrical stills called *cheese boxes*. These could hold up to 1,000 barrels of crude, had greater heat efficiency, and saved on labor costs.

Not long afterward, the first attempts at fractionation were made. Empty towers called *fractionating columns* were placed on the vapor lines from the stills to the condensers. These caught the heavier liquids carried by the vapors and returned them to the still. However, the fractionating pipe still as it is known today did not make its first appearance until 1917.

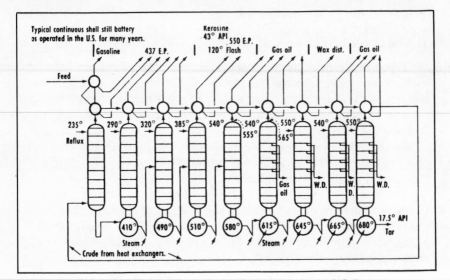

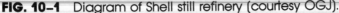

FIG. 10-1 Diagram of Shell still refinery (courtesy OGJ).

The refiners of the first half of the 20th century were faced with a lessening need for kerosene and a growing demand for gasoline. As the demand for product increased, oil men were faced with a choice; either find new fields and use all the crude they could to get the desired end result or develop new processes that would increase the yield of product from existing supplies. As it turned out, they did both. In 1901 Spindletop came in, followed shortly thereafter by the Texas and Oklahoma fields that added millions of barrels of crude to the nation's supply, and refiners discovered new and more productive techniques.

They reasoned that the place to start was with the residuum that was left after the traditional process had taken place. This was a thick, tough deposit left in the bottom of the still. It could be sold for use as fuel in place of coal, but it had to be chopped out by hand.

Dr. William Burton, a Standard Oil of Indiana chemist, developed a successful cracking process that marked a milestone for refiners because now they were able to get a yield of 70 percent distillates of which half was gasoline. The Burton process was in vogue from 1913 to 1920, when it became obsolete (Fig. 10-2).

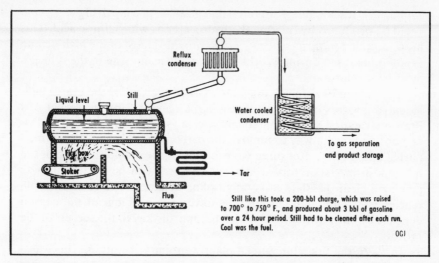

Still like this took a 200-bbl charge, which was raised
to 700° to 750° F., and produced about 3 bbl of gasoline
over a 24 hour period. Still had to be cleaned after each run.
Coal was the fuel.

OGJ

FIG. 10–2 How the Burton process worked (courtesy OGJ).

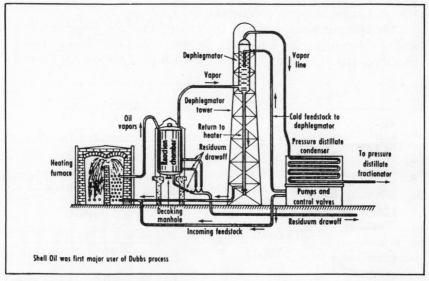

Shell Oil was first major user of Dubbs process

FIG. 10–3 The Dubbs process (courtesy OGJ).

Carbon Petroleum Dubbs developed a cracking process basically
designed to cut down on the coke deposits that had plagued the Burton
process (Fig. 10–3). With the Dubbs process, fractions heavier than gas
oil could be cracked, coke deposits were greatly reduced, and the unit
could run for days without having to be shut down for cleaning—a distinct advantage over other methods, some of which could only operate
for a few hours.

In 1938, researchers discovered that gasoline of a much higher octane could be produced by treating the hydrocarbons with sulfuric acid. This process, called *sulfuric-acid alkylation*, provided most of the high-octane aviation fuel used during World War II and became one of the refiner's most important tools.

The advent of catalytic cracking marked the end of the era of individual innovators in the field of refining. Eugene J. Houdry, owner of a French structural steel firm, sought a catalyst that would crack hydrocarbons and then a way to remove the carbon that subsequently formed during operation. After three years of testing, he found that one of the reactors in his laboratory which was charged with heavy fuel oil was making a clear distillate of good gasoline qualities. The catalyst in the reactor was aluminum silicate. He solved the problem of the carbon deposits by burning them off, which became the key to the success of the process.

In 1936, the first commercial plant went into operation. This was a fixed-bed process using activated bentonite clay. After 10 minutes it would be taken off stream, the vaporized oil sent to another reactor, and the catalyst regenerated with oxygen and gas.

The catalytic process offered the opportunities for providing more and better gasoline components. By the time World War II finally broke out, there were 12 plants in the U.S. providing 132,000 b/d. By the end

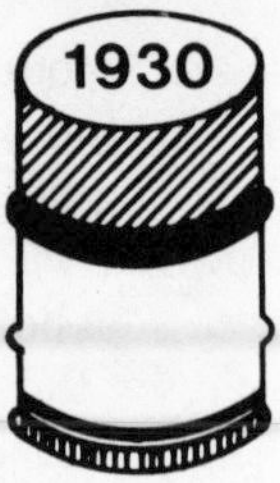

		GALLONS PER BARREL	% YIELD
TODAY	Gasoline	20.8	49.6
	Jet Fuel	2.8	6.6
	Kerosine		
	Gas oil and distillates	8.9	21.2
	Residual fuel oil	4.0	9.3
	Lubricating oils	2.9	7.0
	Other products	2.6	6.3
	Total	42.0	100.0
1930	Gasoline	11.0	26.1
	Kerosine	5.3	12.7
	Gas oil and distillates, residual fuel oil	20.4	48.6
	Lubricating oils	2.4	5.7
	Other products	2.9	6.9
	Total	42.0	100.0

FIG. 10–4 Average product yield from a barrel of crude oil.

of the war, some 34 plants were in operation with a capacity of 500,000 b/d. Cat cracking was the major refining method until 1960 (Fig. 10–4).

The next innovation was *hydrocracking,* which utilized both a catalyst and hydrogen to process residuals or product in the middle-boiling range to high-octane gasoline, jet fuel, and high-grade fuel oil. This came at almost the same time as kerosene reemerged as a demand product. The jet engine had been introduced during World War II, and in the two decades following most piston-engined military and commercial aircraft had been replaced by jets.

Other refining technology today includes *reforming,* which is the use of catalysts and heat to rearrange hydrocarbon molecules without altering their composition. The use of unleaded gasoline in new automobiles is calling for increased refinery capacity and capability for this product. Also in demand are fuel oils of low-sulphur content (Fig. 10–5).

Product	% volume		
	1955	1975	1990
Gasoline	46.4	49.6	33.6
Jet fuel	2.0	6.6	7.8
Kerosene and distillate fuel oil	25.1	21.2	25.0
Residual fuel oil..................	14.7	9.3	19.6
Lubes, wax, coke, and asphalt	6.0	7.0	5.6
Other..........................	5.8	6.3	8.4
Total.........................	100.0	100.0	100.0
Total throughput, 1,000 b/d	7,857	13,224	19,500

FIG. 10–5 Current U.S. refining yield (courtesy OGJ).

BASIS FOR REFINING

Refining is the breaking down of crude oil into the desired products. This is possible because crude is not a single chemical compound but a mixture of hundreds of hydrocarbon compounds, each having its own boiling point. Since there is a whole range of boiling points, when a sample of crude is heated to successively higher temperatures, a boiling point or *crude-distillation curve* (Fig. 10–7) results.

As temperature is raised, a point is reached where boiling starts. This is known as the initial boiling point (IBP). Then boiling continues as temperature continues to be increased. Fig. 10–7 shows that as boiling temperature increases, we first move through the butanes-and-lighter fraction of the crude. This starts a IBP and ends at just below 100°F. The

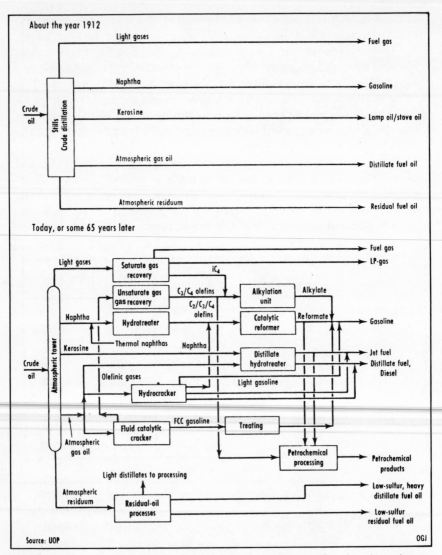

FIG. 10–6 Comparison of early and modern refineries (courtesy OGJ).

fractions boiling through this range are known as the butanes and lighter cut.

Then the next highest fraction or cut starts at just below 100°F. and ends at about 220°F. This is called straight-run gasoline. And starting at 220°F. and continuing to about 320°F. is the naphtha cut. The kerosene cut ranges from about 320°F. to about 450°F., and the gas-oil cut is from about 450°F. to 800°F. The so-called *residue cut* includes everything boiling above 800°F.

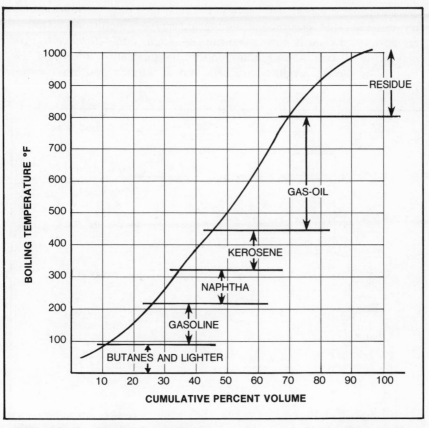

FIG. 10-7 Crude-oil distillation curve and its fractions (after Leffler).

It is not possible to continue the boiling process until all liquid has boiled away. If we heat the residue to temperatures much above 800°F., *cracking* begins to take place. This means that heat breaks down the heavy hydrocarbon molecules in the residue to smaller molecules, hydrogen, and carbon. If we wish to boil the residue to any extent, we must reduce the pressure on the residue by creating a vacuum. Then the residue can be boiled at a lower temperature with no danger of cracking. This process, called *flashing,* will be discussed later.

CRUDE DISTILLATION

If we changed product containers at each cut point, we would be able to recover each cut during the crude-boiling process. We could go through the same process in a large refinery. The process would be called *batch distillation,* and the vessel holding the boiling crude would have to be recharged with fresh crude each time the boiling vessel was emptied.

This process was actually used in the early days of the refining industry, but it became too time-consuming and was very inefficient. So refiners developed what is now called the *continuous crude-distillation process*. Crude is continuously pumped into the distillation tower and, as Fig. 10–8 shows, products are removed at various positions in the tower.

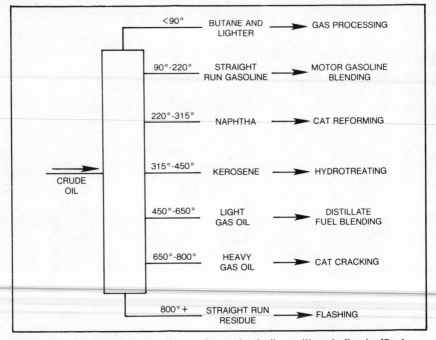

FIG. 10–8 Distilling crude and product disposition (after Leffler).

This is made possible by the characteristics of the *distillation column*. This tall, large-diameter column is a cylindrical hollow steel tower with flat steel trays welded to the column sides every few feet. Slots and holes in the trays permit vapor to flow up the tower and liquid to flow down. Vapor flows through the slots and liquid flows down through a *downcomer,* or pipe leading from the above-mentioned hole in the tray, down to the tray below.

There are several ways to provide orderly flow of vapor through the slots. One of the most common is called a *bubble cap*. A small pipe is welded over a group of slots on the upper surface of the tray. Height of the pipe is equal to the desired level of liquid above the tray when the tower is in operation. The bubble cap, placed on the pipe, has slots in its side to permit vapor to come through the slots, through the pipe, down through the liquid, through the slots, and up through the liquid on the tray.

A downcomer to the tray below is placed opposite the downcomer from the tray above, so that liquid must pass in one direction and then is reversed. Liquid thus flows back and forth across column diameters as it moves down the tower.

When liquid reaches the bottom of the tower, it is removed and sent to a heater. There all but *net bottoms product* (straight-run residue) is vaporized and reentered into the column to provide a continuing source of vapor throughout the column.

When vapor reaches the top of the column, it is removed and all but the *net overhead product* (butane and lighter) is condensed and put back into the tower as a continuing source of liquid throughout the column.

The continuous admixture of liquid and vapor throughout the tower establishes a temperature pattern such that the tower-tip temperature corresponds to the upper cut point for butanes and lighter. The tower-bottom temperature is the upper cut point for gas oil. The cut points for the intermediate fractions may then be found with the lighter cuts in the top portion of the tower and the heavier cuts in the bottom portion.

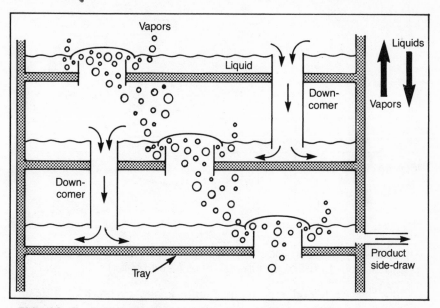

FIG. 10-9 Downcomers, bubble-cap trays, and side draws (after Leffler).

At the appropriate tower locations, depending on temperatures, the intermediate cuts are withdrawn from the tower through what are known as *side draws* (Fig. 10-9). Each of the cuts shown in Fig. 10-8 is further processed in other parts of the refinery.

FLASHING

The liquid recovered from the very bottom of the crude tower is subjected to vacuum flashing. As stated earlier, further boiling would lead to cracking or thermal breakdown of the very heavy residue. So a physical chemistry phenomenon is called into play: as pressure decreases, boiling temperature for any given liquid also decreases.

If the effective tower operating pressure is lowered, then the residue could be heated further at a lower temperature without cracking. This is accomplished by lowering the pressure, or operating at a vacuum.

Atmospheric pressure is about 14.7 psi (pounds per square inch), and this is the approximate pressure on the crude-distillation tower. The pressure in the flasher is about 5 psi, established by a vacuum pump at the top of the vessel (Fig. 10–10).

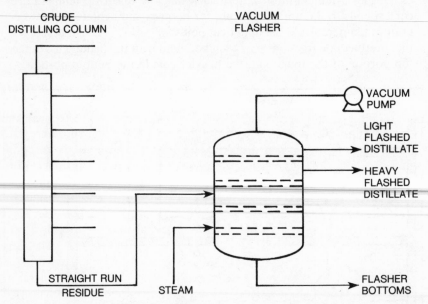

FIG. 10–10 Vacuum flashing allows the refiner to recover more products from the straight-run residuum (after Leffler).

It takes heat to boil or flash liquids into vapors, so superheated steam is introduced into the flasher. This also adjusts the partial pressure of hydrocarbons in the vessel and allows close pressure control.

Several streams can be taken off the flasher: light flashed distillate, heavy flashed distillate, and flasher bottoms. The distillate streams go to other refinery locations for further processing or fuel-oil blending. The flasher bottoms may be used in residual fuel, may be blended to asphalt, or may go to a thermal cracker where the big hydrocarbon molecules are broken down to make thermally cracked stock for still further processing.

So we have now separated from the crude several cuts in the distillation column and some more in the flasher. These cuts are now ready for further processing and blending into refined products.

CATALYTIC CRACKING

The two cuts just lighter than straight-run residue from the crude-distillation column are light and heavy gas oils (Fig. 10–8). Prime use of the heavy gas oil especially is as a feedstock for the catalytic cracking unit. Light gas oil may also be fed to this unit, either as a separate stream or as a mixture with heavy gas oil or it may be blended with other stocks into a heavy-distillate fuel oil. Light flashed distillate from the vacuum flasher may also go to the cat cracker.

In the United States, most gas oils are fed to the cat cracker for gasoline production. In other countries, the bulk of the gas oils goes to distillate fuel oil. The difference is that the U.S. with its large car population has a healthy appetite for gasoline production; the cat cracker, above all else, is a gasoline maker.

In the catalytic-cracking process, gas oil is subjected to heat and pressure in the presence of a catalyst. A catalyst is a substance that causes or enhances a reaction, but which is not changed in the reaction. In the cat-cracking process, the catalyst is developed and selected to convert the gas oil largely to gasoline, though the stream from the reactor contains a full range of hydrocarbons, methane through residue.

The heart of the cat cracker is the reaction chamber or reactor. Field gas oil is heated to about 900°F., mixed with a catalyst (usually a very fine powder), and introduced into the reactor (Fig. 10–11). The time required for reaction is only a few seconds. Then the spent or used catalyst is separated from the hydrocarbon and sent to the regenerator. Here it contacts air and the carbon deposited during the reaction process is burned off at about 1,100°F. under carefully controlled conditions.

Fresh (regenerated) catalyst exits from the bottom of the regenerator and is again joined with more incoming feed before entering the reactor. As the result of the combustion of carbon in the regenerator, flue gas from the regenerator is a mixture of carbon dioxide and carbon monoxide. Due to the heat released in the burning, this stream is very hot, and some heat is generally recovered in some other parts of the cracking process.

The cracked product from the reactor is pumped into a fractionator where five streams are recovered: C_4 and lighter, cat gasoline, cat light gas oil (LGO), cat heavy gas oil (HGO), and cycle oil. The overhead light stream goes to the cracked gas plant (see next section). Cut gasoline

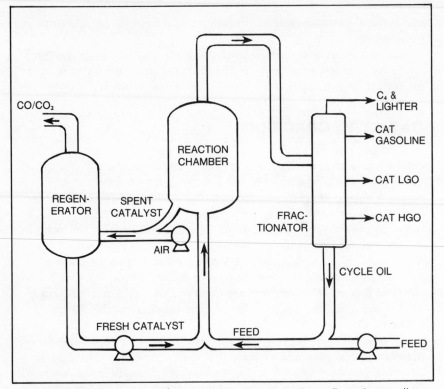

FIG. 10-11 Cat-cracking unit adds greatly to the refinery's gasoline yield (after Leffler).

goes to gasoline blending (see a later section). The cat-cracked LGO goes to light-distillate fuel oil blending, while HGO goes to heavy-distillate fuel oil. Cycle oil usually is sent back to the reactor feed; thus, it is recycled to extinction.

During the summer, some of the lightest LGO is moved over to gasoline (gasoline mode). In winter, some of the heaviest gasoline is moved into the LGO to maximize fuel-oil production (heating-oil mode).

Some typical cat-cracker yields are shown in the table below:

		% volume
Feed:	Heavy gas oil	40.0
	Flasher tops	60.0
	Cycle oil	(10.0)*
		100.0
Yield:	Coke	8.0
	C_4 and lighter	35.0
	Cat-cracked gasoline	55.0

Cat-cracked light gas oil	12.0
Cat-cracked heavy gas oil	8.0
Cycle oil	(10.0)*
	118.0

*Recycle stream not included in feed or yield total

REFINERY GAS PLANTS

Almost all refinery processing units generate some gas (butanes and lighter). This gas is handled in a *gas plant*. The bulk of refining gas is saturated (no olefins), the principal stream of this type being the crude distillation overhead (Fig. 10–12). The saturated gas is collected in the *sats gas plant* for further processing. Those streams containing olefins or *unsaturates* go to the *cracked-gas plant* for further processing.

Sats Gas Plant

Saturated gases from various parts of the refinery are collected at the sats gas plant. They are under very low pressure and must be compressed to higher pressure for processing (Fig. 10–12). Following compression

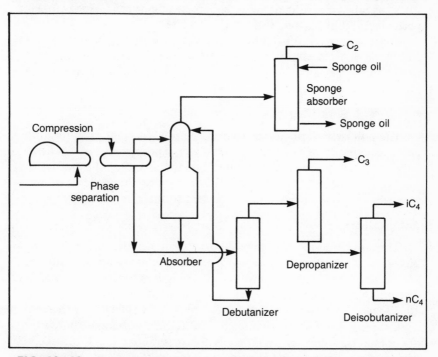

FIG. 10–12 Sats gas plant recovers light hydrocarbon products (after Leffler).

and cooling, some of the gas is liquefied and the two-phase mixture is sent to a gas-liquid separation drum. Overhead gas goes to an absorption system, while the liquid bottoms go to a fractionation train for separation into several products.

In the processing scheme, liquid from phase separation is joined by rich absorption oil from the absorber processing the gas stream. This brings in propane and heavier liquids, absorbed from the gas stream. The combined liquids feed a debutanizer fractionation column in which all butanes and lighter are taken overhead. The bottoms stream is recirculated to the absorber as lean oil.

The overhead stream from the debutanizer feeds a depropanizer, from which a propane product is taken overhead. The bottom stream from the depropanizer feeds a deisobutanizer, which makes an isobutane overhead product and a butane bottom product.

The vapor from phase separation goes through the aforementioned absorber for removal of propane and heavier in the absorption oil, which then delivers these materials to the liquids fractionation train. Gas from the absorber goes through the sponge absorber for removal of lean oil lost from the absorber column.

Product gas from the sponge absorber contains methane and ethane. These usually go to refinery fuel, though the ethane can be recovered for petrochemical processing if desired.

Propane recovered from the depropanizer goes largely to LPG use. Isobutane goes to alkylation (see next section), and normal butane is used almost entirely as a gasoline-blending component for octane improvement and vapor-pressure control.

Cracked Gas Plant

The cracked gas plant is similar to the sats gas plant except that the gases contain olefins or unsaturates: ethylene, propylene, and butylenes. Ethylene usually goes with methane and ethane to the refinery fuel system, though it can be recovered for petrochemical processing if desired. Propylene and butylenes usually are sent to the alkylation plant.

Alkylation

The alkylation reaction joins an unsaturate with isobutane to form a high-octane component in the gasoline range. Propylene and butylenes are the unsaturates used. The reaction requires a catalyst and either hydrofluoric acid or sulfuric acid may be used for this purpose. Fig. 10–13 shows the flow diagram for a sulfuric-acid alkylation process.

Cracked gases, fresh isobutane, and recycle isobutane (unreacted) are joined in a chiller, which cools the mixture to the 40°F. reaction temperature. Then the feed liquids are joined with the acid in the reactor system. A residence time of about 20 minutes is needed, so a battery of

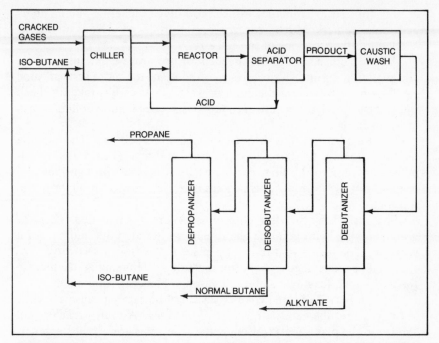

FIG. 10-13 Alkylation plant boosts gasoline yield, octane (after Leffler)

large reactors is used. Mixers ensure intimate contact of feeds and acid.

After sufficient residence time, the mixture moves to the acid separator. Here, there is no agitation, and the heavier acid quickly settles out and is withdrawn for return to the reactor. Hydrocarbon liquid from the acid separator is caustic washed to remove any remaining traces of acid.

Then the product stream is passed through a fractionation train to recover alkylate, normal butane, isobutane, and propane streams. Butane and propane are handled as before (sats gas plant), and isobutane goes back to the chiller for another pass.

Catalytic Reforming

The catalytic reforming process operates on naphtha from the crude tower. The feed may also contain minor amounts of naphthas from coking, thermal cracking, and hydrocracking operations (see later sections). Unlike the cracking processes where large molecules are whittled down to smaller ones, the cat-reforming process merely rearranges the naphtha-sized molecules without actually breaking them down.

Typically, there are high concentrations of paraffins and naphthenes in the feed naphtha. The cat reformer causes many of these materials to be transformed to aromatics, which have much higher octane numbers and some isomers. Much hydrogen is also produced. This is utilized as needed in other sections of the refinery.

A rather unusual catalyst is needed for this process. Alumina, silica, and platinum go into this catalyst, and some of the more sophisticated catalysts have the metal rhemium as well. Great care is taken to keep track of this catalyst because each unit has several million dollars' worth in its inventory.

Fig. 10–14 shows flow through a typical fixed-bed unit. The naphtha feed is pressured to 200–500 psi pressure and heated to 900–975°F. It is then charged to the first reactor, where it trickles through the catalyst and out the bottom. This process is repeated twice in the next two reactors.

Product is then run through a cooler where it is liquefied. Then the accompanying hydrogen-rich gas stream is separated out and part of it recycled. The rest is sent to the gas plant for hydrogen recovery.

Liquid product feeds a stabilizer that takes off butanes and lighter (to the gas plant), leaving a stabilized gasoline-blending component.

After a period of time, coke deposits on the catalyst cause a decline in activity and the catalyst must be *regenerated*. A fourth reactor (not shown on the plan sheet) is used, and one regenerator is being regenerated at all times with three onstream.

After two to three years, the catalyst has collapsed enough to cause sufficient activity drop. The old catalyst must then be removed and new catalyst substituted for it.

RESIDUE REDUCTION

The processing units outlined thus far tend to concentrate on midboiling-range materials and to leave comparatively large volumes of heavy materials. Basically, two processes are used to process the heavy materials and convert appreciable amounts of residuals to lighter materials: thermal cracking and coking.

Thermal Cracking

Feed to a thermal-cracking unit is usually flasher bottoms, although cat-cracked heavy gas oil and cat-cracked cycle oil may also be used.

If a broad range of feeds is to be processed, the lighter materials are kept separate from the heavier stocks. Each is fed to a separate furnace, since temperature requirements to cause cracking are higher for the light products. The furnaces heat the feed to the 950–1,020°F. range (Fig. 10–15). Residence time in the furnaces is kept short to prevent appreciable coking in the furnace tubes. The heated feed is charged to a reac-

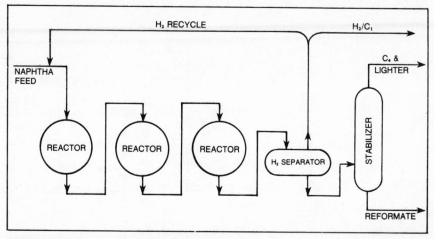

FIG. 10–14 Catalytic reforming boosts naphtha octane greatly (after Leffler).

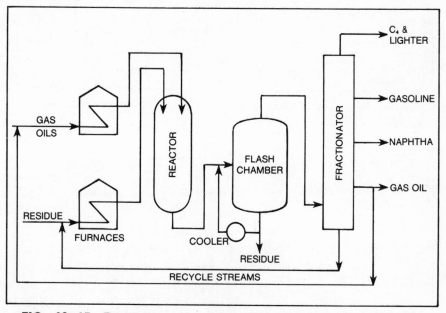

FIG. 10–15 Thermal cracker helps reduce residual yield (after Leffler).

tion chamber, which is kept at high enough pressure (140 psi) to permit cracking but not coking.

Reactor effluent is mixed with a cooler recycle stream, and cracking is then stopped. The combined stream goes to a flash chamber, and the lighter materials flash overhead to a fractionator recovering light prod-

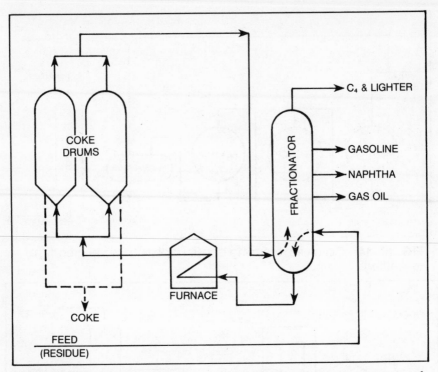

FIG. 10-16 Coking reduces residual yield, enhances recovery of lighter materials (after Leffler).

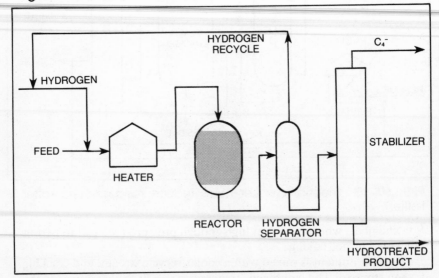

FIG. 10-17 Hydrotreater removes sulfur, nitrogen, and heavy metals from refinery streams (after Leffler).

ucts: butanes and lighter (goes to the gas plant), gasoline (goes to gasoline blending), naphtha (to catalytic reforming), and gas oil (to cat cracker). The net heavy residue usually is blended into residual fuel.

Coking

Coking is severe thermal cracking. The feed is heated at high velocities to about 1,000°F. and charged to a coke drum, where the actual cracking takes place (Fig. 10–16). The lighter cracked product rises to the top of the drum and is drawn off. Heavier product remains and cracks to coke, a solid coal-like substance.

Vapors from the top of the drum are sent to the fractionator, which produces butanes and lighter, gasoline, naphtha, and gas-oil streams.

When the drum is full of coke, the vessel is taken offstream, cooled, and opened up. Decoking usually is accomplished by a high-pressure water jet that delivers lumps of coke to trucks or railway cars for shipment.

HYDROTREATING AND HYDROCRACKING HYDROGEN

Many refinery streams have sulfur, nitrogen, and heavy metals in them. Treating with hydrogen, or *hydrotreating,* can effectively remove these materials and bring about some other benefits as well. In the hydrotreating process, the stream to be treated is mixed with hydrogen and heated to 500–800°F. (Fig. 10–17). The hydrogen-oil mixture is then charged to a reactor filled with pelleted catalyst. Several reactions take place:

1. The hydrogen combines with sulfur to form hydrogen sulfide (H_2S).
2. Nitrogen in some of the nitrogen compounds is converted to ammonia.
3. Any metals entrained in the oil are deposited on the catalyst.
4. Some of the olefins, aromatics, and naphthenes get hydrogen saturated and some cracking takes place, causing formation of butanes, propane, and lighter.

The stream from the reactor goes to the hydrogen separator, from which hydrogen is recycled to the reactor. The remaining materials go to a stabilizer where light ends, including propane and lighter, hydrogen sulfide, and a small amount of ammonia are taken overhead. Hydrotreated product exits at the bottom of the tower.

Many of the refinery intermediate streams that require further processing are hydrotreated. Some of the product streams are treated as well. Residual fuel, jet fuel, kerosene, and light and heavy distillate fuels are often hydrotreated.

Hydrocracking increases the overall refinery yield of quality gasoline-blending components. The process really is catalytic cracking in the presence of hydrogen. It can take as feed low-quality gas oils that otherwise would be blended into distillate fuel. The hydrocracker thus permits a fairly wide swing in a refinery's operation, from ultrahigh yield of gasoline in summer to high yield of fuel oil in the winter.

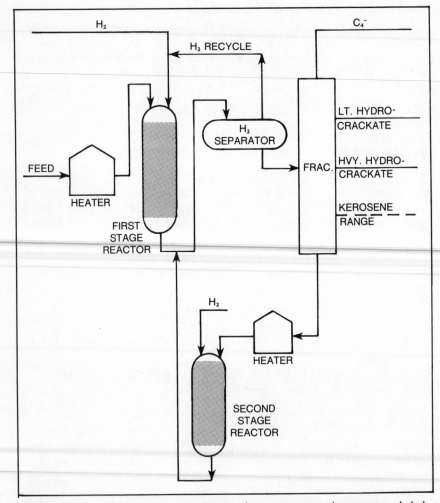

FIG. 10-18 Two-stage hydrocracker process increases total amount of high-quality blending components for gasoline (after Leffler).

The hydrocracking process features fixed-bed reactors—two or more (Fig. 10–18). Feed is mixed with hydrogen vapor, heated to 550–750°F., pressurized to 1,200–2,000 psi, and charged to the first-stage reactor. Here, about 40–50% of the feed is cracked to gasoline-range material.

The stream from the first-stage reactor is cooled, liquefied, and run through a separator. Hydrogen is recycled, while the liquid is charged to a fractionator. Here, such streams as butanes and lighter, light hydrocrackate, heavy hydrocrackate, and kerosene are taken off. The bottoms stream then goes to the second-stage reactor.

Conditions in the second stage include higher temperature and pressure. Outlet stream is sent to the hydrogen separator and thence to the fractionator. The product gasoline-blending components are the light and heavy hydrocrackates.

The hydrocracker, in addition to providing high-quality products, also has a 20–25% volume gain over the feed, shown in the table below:

	Volume balance
Feed	
Coker gas oil	0.60
Cat-cracked light gas oil	0.40
	1.00
Product	
Propane	
Isobutane	0.02
Normal butane	0.08
Light hydrocrackate	0.21
Heavy hydrocrackate	0.73
Kerosene range	0.17
	1.21

Hydrogen

The normal source of hydrogen is the cat reformer. The light-ends stream from the reformer column *(stabilizer)* is deethanized to produce a high-concentration hydrogen stream (see cat-reforming flow sheet). However, the refinery may require more hydrogen than the cat reformer produces. Then a stream-methane reformer will be installed. As indicated, methane (the chief component of natural gas) and water are the two feedstocks required.

First, methane and water react at about 1,500°F. to form carbon monoxide and hydrogen. Then more water reacts with the carbon monoxide to give carbon dioxide and hydrogen. Next, a solvent extraction process removes all but traces of the carbon dioxide. Finally, in a reaction called *methanation,* the remaining traces of carbon monoxide and carbon dioxide are removed by conversion back to methane and water.

GASOLINE BLENDING

Vapor Pressure

The discussions on refinery processing have indicated a number of components available for gasoline blending: normal butane, reformate (94 RON), reformate (100 RON), light hydrocrackate, heavy hydro-crackate, alkylate, straight-run gasoline, straight-run naphtha, cat-cracked gasoline, and coker gasoline (RON stands for *research octane number*). All of these must be blended together to give the required vapor pressure, give the required octane number, and use up all of the available stocks.

The vapor pressure of gasoline is that pressure it exhibits to vaporize. Each component has a different vapor pressure, and for all practical purposes gasoline-blending components blend linearly by vapor pressure.

Required vapor pressure, related to the ability of gasoline to vaporize in an automobile carburetor, varies from one place to another and in summer and winter. And blending to vapor pressure is a function of ambient temperature in any one location. The practice is to blend various grades of gasoline according to octane-number requirements (discussed later in this section). Then normal butane is added as necessary.

Octane Number

Blending for octane number in the refinery again calls for a knowledge of amounts of various available stocks and their octane numbers. *Octane number* is a measure of the knocking characteristics of the gasoline. The higher the octane number, the less tendency an automobile engine burning the gasoline will have to knock.

The term octane number comes from the test used to determine gasoline knocking characteristics. A test engine mixes two standard fuels and tests the mixture against a given product gasoline. One of these two fuels is iso-octane (100 octane); the other is normal heptane (0 octane). Two octane numbers are determined on a gasoline blend: the research octane number (RON) and the motor octane number (MON). The *research octane number* simulates driving under mild conditions, while the *motor octane number* simulates driving operations under load or at high speed. The number the FTC requires to be posted at the service-station pump is (RON + MON)/2.

One other factor that needs to be covered in gasoline blending for octane is the use of tetraethyl lead (TEL). Often, the refiner is faced with an octane deficiency when he considers the various grades of gasoline that he must make. Until 1974, he was free to use TEL as needed (up to 3 cc/gal) to suppress knock and thus enhance octane number.

There never has been any definite proof that burning TEL in a gasoline engine causes any detriment to human health. True, when appreciable amounts of lead get into the bloodstream, health problems occur. But there is no evidence that the lead oxide that leaves in engine exhaust has ever, or could, get into a person's bloodstream.

Nevertheless, in 1974 EPA mandated a gradual phasedown of lead content in gasoline, starting in 1975. The result today is that many smaller refiners cannot meet octane-number specifications and some larger plants are finding it increasingly difficult to do so.

With several blending stocks exhibiting different octane numbers and vapor pressures, making gasoline blends as required to meet specifications for several gasoline products is very difficult to work out. Probably the most successful technique for coping with all these variables (plus using up all available stock) is linear programming on a large computer to simulate refinery operations. The linear-programming technique solves all these requirements at maximum profit.

DISTILLATE FUELS

Distillate fuels are those blended from light gas-oil-range streams. Diesel oil and furnace oil are the two most referred to. However, in many refineries, these are one and the same, sold out of the same tank.

There are several grades of diesel fuel. Regular diesel runs about 40–45 cetane; premium runs 45–50. The *cetane number* is determined much like octane number for gasoline. The test blends for comparison are mixtures of cetane (100) and alphamethylnaphthalene (0). All light gas oils are candidates for diesel-fuel blending. Straight-run gas oil is prime, with cetane number of 50–55.

The gas-oil hydrocarbons can also be blended into furnace oil, the most popular petroleum heating oil. It has a higher heating value than the lighter hydrocarbons, and its ignition characteristics are safer. Furnace oil also is easier to handle and more pollution-free than residual fuels. Furnace oil is called several other names as well (2 fuel, distillate fuel, two oil).

Flash-point and pour-point specifications are most important considerations for furnace oil. The *flash point* is the lowest temperature at which enough vapor is given off to form a combustible mixture with air. This must be high enough to preclude danger of fire in the heating system.

RESIDUAL FUELS AND ASPHALT

The refiner has two alternatives to handle the very heaviest high-boiling materials (residue) in the crude: residual fuel or asphalt. When asphaltenes occur in the residue, a strong, stable asphalt can be made. Asphalts are characterized in four ways: straight run, blown, cutbacks, and emulsions (Fig 10–19).

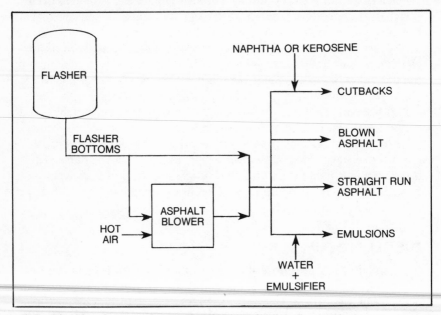

FIG. 10–19 Options in production asphalt (after Leffler).

Straight-run asphalts come from deep flashing of crude. Two important tests for these asphalts are *softening point* and *penetration*. Softening point is that temperature at which a standardized object will start to sink in the asphalt. Penetration refers to applied asphalt and measures depth of penetration of a needle at standard test conditions.

Consistencies of softer grades of asphalt can be changed by blowing with hot air, causing a chemical reaction. The resultant asphalt product is harder and more rubbery.

To reduce the severity of application conditions, a thinner (cutback) may be added. Naphtha or kerosene is usually used. After application, the diluent will evaporate, leaving a hard, durable asphalt.

Emulsion asphalts are also used to facilitate handling during application. The asphalt is emulsified with water and applied. Then the water evaporates, once again leaving a hard, durable asphalt.

Crude residue may also be used as residual fuel. It must be heated at all handling points to prevent solidification. Specifications include viscosity (how thick and sticky it is), sulfur content, and flash point.

Generally, flasher bottoms (crude residue) must have some kind of diluent to meet the maximum viscosity specifications. Cat-cracked heavy gas oil, relatively low in viscosity and of low value as fuel for conversion processes, usually is used.

Residual fuels usually have some sulfur and may have a great deal. The high-sulfur resids often are blended off to meet sulfur specifications, usually 1% sulfur maximum.

Flash point is important because the residual fuel must be heated to flow. And some relatively high flash-point material may have ended up in the resid. Usually, flash point is the limiting factor as to what can be dumped into the resid.

CHAPTER 11

Petrochemicals

There is more in a barrel of crude oil than gasoline for your car. More, in fact, than the other traditional petroleum products such as oil, grease, wax, and asphalt. Four percent of the natural gas and crude oil in the U.S. is used as feedstock, or raw materials, by chemical companies to make products valued at $40 billion annually. Thus, a single barrel of crude may be turned into consumer products worth almost $300.

HOW THE INDUSTRY BEGAN

The petrochemical industry has grown hand in hand with the refining industry. The first large-scale advance for both came in about 1912 when cracking made its first large-scale appearance, and both refiners and petrochemical manufacturers have benefitted from each new advance in cracking methods. Cracking not only improved both the quality of gasoline and the amount produced from each barrel of feedstock, it also vastly increased the output of alkenes. These are highly reactive unsaturated hydrocarbons that form the basic material for the petrochemical industry.

Petroleum is composed mainly of compounds of the elements carbon and hydrogen or, more simply, hydrocarbons. But nitrogen and sulfur are also found in crude oil. All four of these are extremely valuable in the manufacture of chemicals. Since these chemicals are derived from petroleum, they are called petrochemicals.

THE STRUCTURE OF HYDROCARBONS

Hydrogen and carbon both have their own distinctive atomic structures. It is by rearranging their individual molecular structures that chemists and chemical engineers are able to produce many industrial materials. Hydrogen has the simplest atomic structure of any element. It has no neutrons in its nucleus, but instead has a single electron moving around it at a very low energy level. There is room for only one additional electron. An atom becomes chemically stable when its outermost energy level is filled. Since hydrogen has only one energy level with a single vacancy, it becomes stable when one more electron is acquired.

By contrast, the carbon atom is much more intricate. There are six neutrons and six protons in the nucleus. Two electrons move in the innermost and lowest energy level and four in the next. This second energy level has room for eight electrons, so four more must be added for stability. Hydrogen and carbon then achieve stability and form hydrocarbons sharing electron pairs. These shared electrons jointly occupy the outer levels of both atoms. Such sharing is called *covalent bonding* (11–1).

Large hydrocarbon molecules such as those found in petroleum are formed by the bonding of many hydrogen and carbon atoms. The properties of hydrocarbons thus depend upon the number and arrangement of the hydrogen and carbon atoms in their molecules. Since compounds with similar structures have similar properties, the chemist can aim his research at group rather than individual characteristics.

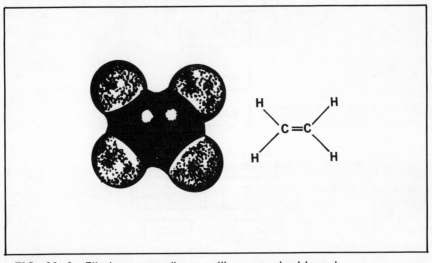

FIG. 11–1 Ethylene, an alkene with a covalent bond.

Alkanes

One important group derived from petroleum is the alkane or paraffin homologenes series (Fig. 11–2). It includes the gases methane (CH_4), ethane (C_2H_6), propane (C_3H_8), and butane (C_4H_{10}). Also included are the liquids pentane (C_5H_{12}) and hexane (C_6H_{14}), as well as the solid heptadecane ($C_{17}H_{36}$).

The so-called normal alkanes are those in which the carbon atoms are strung in a single chain. Other hydrocarbons whose carbon atoms branch out are called *branched-chain* (Figs. 11–3 through 5).

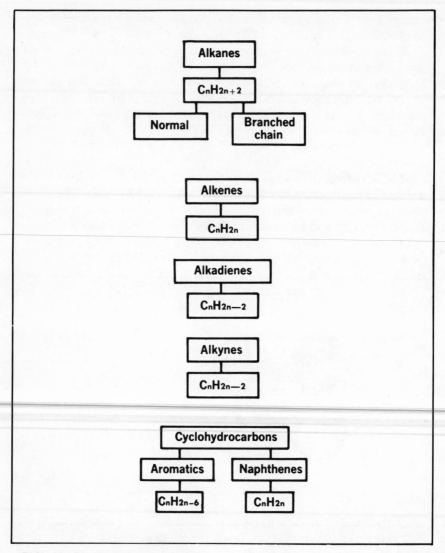

FIG. 11-2 Chemical families.

Isomerism

The alkanes are compounds with a common chemical formula, but because their structure is different they have different melting and combustion temperatures. Such compounds are called *isomers*. Isomers have the same chemical formula and molecular weight but different structures. The different structures lead to different physical and chemical properties. With each additional carbon atom, the number of possible

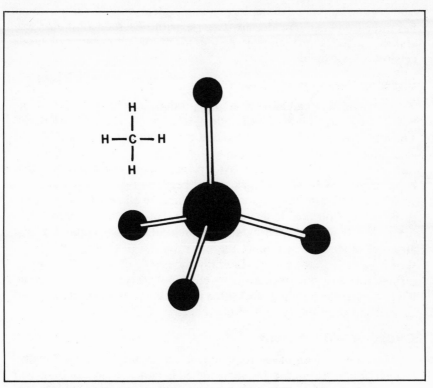

FIG. 11–3 Methane molecule—simplest of the hydrocarbons.

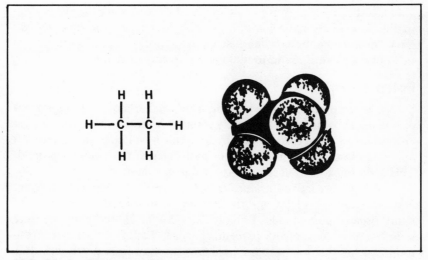

FIG. 11–4 Ethane, a straight-chain molecule.

isomers increases. For example, pentane (C_5H_{36}) with 17 carbon atoms has three isomers. Heptadecane ($C_{17}H_{36}$) with 17 carbon atoms has several thousand. Isomerism plays an important role in making new products out of petroleum.

Unsaturated Hydrocarbons

Petroleum also can be processed to produce other types of hydrocarbons. These are the alkenes (olefins), such as ethene or ethylene, in which the molecule has a double bond consisting of two pairs of shared electrons. Another is the alkadienes (diolefins) such as butadiene. A third is the alkynes, such as ethyne or acetylene, which have a triple bond of three pairs of electrons shared between two carbon atoms.

These three are much more reactive than the alkanes because of the carbon-to-carbon double and triple bonds that permit additional reactions. The bonds open easily and become attachment points for new atoms of other elements. Since there is room for these additions, these series are called *unsaturated*. Literally thousands of derivatives can be formed this way, and over the years ethylene, acetylene, and butadiene have become the building blocks that have led to dozens of useful products and spurred the growth of the industry.

Cyclic Hydrocarbons

Yet another type of hydrocarbons is the cyclics, so called because instead of their carbon atoms being linked in either straight or branched chains they have a ring of carbon atoms in their structure (Fig. 11–6). Benzene and its derivatives form the group with the greatest economic value. Collectively they are known as the *aromatics* since they were originally found in aromatic gums or oils. (The others are known as *aliphatics*.) These compounds when first manufactured were made from coal tar, but ways were found to make them from petroleum.

Polymerization

Some compounds can be built up from smaller molecules and others can be made by breaking down larger ones. As noted above, the double and triple bonds of unsaturated hydrocarbons will open up under suitable conditions and join with other molecules of the same compound. This combining of molecules is known as *polymerization*.

The highly reactive unsaturates, which are byproducts of cracking, and especially the olefins are the starting point for the manufacture of many organic compounds. Prior to World War II, the petrochemical industry provided only 45 percent of these. Today, 70 percent of the total U.S. production of organic chemicals comes from petroleum. It is estimated that a half million or more compounds could be synthesized from the hydrocarbons in natural gas and crude oil.

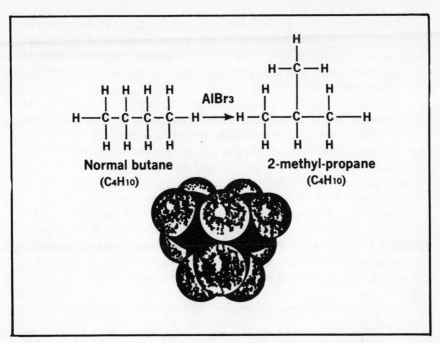

FIG. 11–5 Methyl-propane, a branched-chain hydrocarbon.

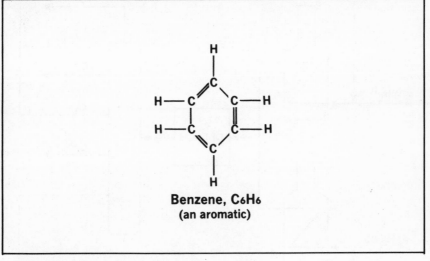

Benzene, C₆H₆
(an aromatic)

FIG. 11–6 Benzene, a cyclic hydrocarbon.

ETHYLENE PRODUCTION

One of the general types of petrochemical processing directly related to petroleum is ethylene production. This serves as a branch of the more diversified petrochemical industry.

The ethylene or olefins plant can take ethane, ethane/propane mix, propane, butane, naphtha, or gas oil as feedstock. Those plants feeding the heavier materials are more complicated, however, and we will examine only the ethane/propane cracking process to demonstrate the fundamentals.

As shown in Fig. 11–7, ethane and propane can be fed separately or as a mixture to the cracking furnaces where short residence time followed by a sudden quench yields a high volume of ethylene. The feed is not completely reacted, so downstream in the product fractionator the remaining ethane and propane are split out and recycled to the feed. Ethane generally is recycled to extinction, but some of the propane goes with the propylene.

The plants that crack the heavier liquids—naphtha and gas oil—create ethane on a once-through basis, so those olefin plants often have a furnace designed to handle the recycled ethane. As a result of the cracking reaction that produces the olefins, some butadiene is also formed. It is used in making plastics and rubber compounds.

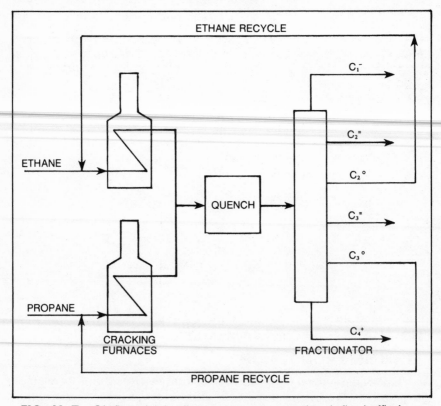

FIG. 11–7 Olefins plant: ethane-propane cracker (after Leffler).

SOLVENT RECOVERY OF AROMATICS

The second general type of petrochemical processing directly related to petroleum is BTX recovery. The aromatics compounds (benzene, toluene, and xylene) have many chemical applications. These three compounds appear in relatively large quantities in catalytic reformates (see chapter 10). So these materials are recovered by treating the reformate with a solvent that preferentially dissolves them. The solvent with the dissolved aromatics can then be readily separated from the rest of the compounds.

The dissolved benzene, toluene, and xylenes (or BTX) can be easily recovered by a simple distillation of the solvent. Then the BTX materials may be separated into individual components by further distillation.

Fig. 11–8 shows how the solvent-recovery process works. A *heart cut* reformate is the feed. This is obtained by separating off materials lighter

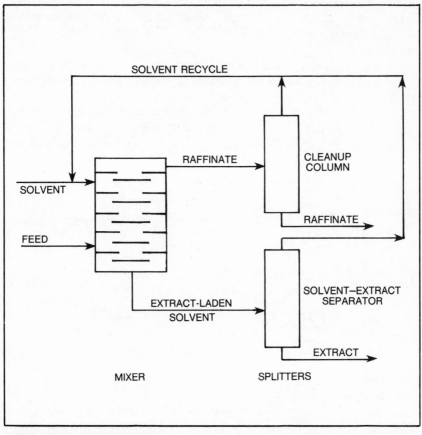

FIG. 11–8 Solvent recovery process yields aromatics for chemical use (after Leffler).

than benzene and materials heavier than xylenes. Feed and solvent enter a mixer column where solvent flows downward, contacting the rising feed stream. The solvent absorbs BTX as it proceeds down the column.

The overhead raffinate contains almost no BTX. It is treated for solvent removal (which is recycled) and returned to the reformer. The bottoms extract-laden solvent moves to the solvent-extract distillation separator. Recycle solvent is taken overhead, and extract containing product BTX is ready for fractionation. Some of the solvents used are sulfolene, phenol, acetonitrile, and liquid SO_2.

Alcohol

Acohol is a hydrocarbon in which one or more of the original hydrogens has been replaced by a hydroxyl group. For many years, one of the most common alcohols, ethyl or grain alcohol, was made by the yeast fermentation of starches and sugars. But because the demand for ethanol as a solvent increased drastically, now 90 percent is produced from ethylene.

The same is true for methanol, which was formerly distilled from wood and is now made from methane. Common rubbing alcohol, isopropyl, is produced from propane.

Some alcohols contain two or more of the functional hydroxl groups. When two hydrogen atoms are replaced by hydroxl groups, the compound becomes a glycol. Ethylene glycol made from ethene is used as an antifreeze and to synthesize Dacron.

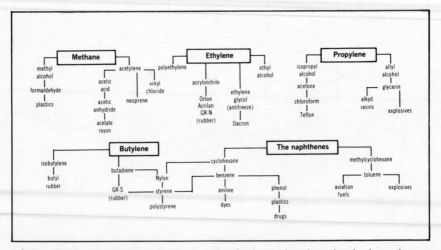

FIG. 11-9 Basic products made from petrochemicals (courtesy Chemistry and Petroleum).

Glycerin, which used to be a byproduct of the soap industry, is now made from petroleum and is used in the explosive, nitroglycerin.

Detergent

A detergent is a cleaning agent consisting of molecules with two distinct functions. The hydrocarbon end is fat soluble and the other end is electrically charged and water soluble. These distribute themselves where the oil and water meet, reducing the surface tension of both so that dirt particles are easily suspended to form an emulsion that is washed away with rinse water. Petroleum-derived detergents are much more effective than soap.

Polymers

In polymerization, two or more smaller molecules, usually from unsaturated compounds, are combined to form one larger molecule. The

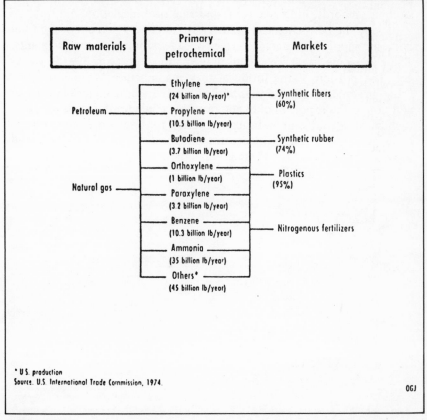

FIG. 11-10 Percentage of petrochemical use (courtesy OGJ).

reaction conditions can be controlled to produce polymers of varying size. As the size and weight of the molecules increase, the melting point of the polymers rises and the solubility decreases. This way, polymers with many different physical properties can be made. Fibers are made from polymers that can be stretched out to form a filament. Those that have rubber-like qualities are *elastomers* or synthetic rubber and others are *plastics*. A few of the products of polymerization would include polyethylene plastics, nylon, and neoprene synthetic rubber.

Petrochemicals Today

These are but a few of the products possible from petrochemicals—a complete listing of consumer products from aspirin to fertilizers, raincoats, and telephones would fill a volume by itself. Many products were discovered and synthesized by research chemists before there was even a market or demand for them. This is true of ethylene glycol before it was put in wide use as antifreeze, and vinyl before it gained wide usage as clothing, records, inflatable toys, and electrical insulation.

Today the petrochemical industry utilizes 8 to 9 percent of the total domestic supply of oil and gas as feedstock and for fuel to operate plants. There are some 1,000 such plants in the nation employing 390,000 workers. The portion of the products they manufacture for export alone brings in some $5 billion a year to the United States' balance of trade. In addition to the export market and products made for domestic use, petrochemical manufacturers furnish materials for final processing to other manufacturers. These products are valued at $279 billion a year and provide employment for 11 million people.

CHAPTER 12

Instrumentation and Controls

The various types of equipment described in this chapter are related to the control and measurement of a liquid or gas. The purpose of this equipment is to maintain or measure the liquid or gas at a specific pressure, temperature, flow rate, volume, or liquid level. For simplicity, liquid and gas will often be referred to collectively as *controlled medium.*

Final control is accomplished either directly or indirectly by decreasing or increasing the size of the opening through which the controlled medium is passing. The predominant measurement in the refining and processing industry is flow rate because it is a continuous liquid flow operation in which monitoring and controlling are essential. Flowmeters are most often used and monitored for this purpose.

In the natural-gas industry, flow control of gas is important. Control of gas-system flow requires regulating pressure. Gas pressure will vary with temperature and volume. Control of gas-system pressures is usually accomplished with pressure-reduction equipment. Control is often accomplished with a gas pressure regulator; however, it could be done manually with a globe valve or any other type valve adapted to throttling or close control.

Pressure is a common term in the gas industry. It can be any force from less than atmospheric (vacuum) to thousands of pounds above atmospheric pressure.

TEMPERATURE MEASUREMENT

Temperature measurement is common in refining and processing. It is one of the first and oldest measurements associated with the industry. Temperature is a measurement of the hotness or coldness of a substance. The *thermometer* is an instrument used for determining the temperature of a body or space. There are several types of thermometers used for this purpose. Common thermometers consist essentially of a confined sub-

205

stance such as mercury, the volume of which changes with a change in temperature. The clinical thermometer is one example in which body temperature causes the confined liquid to rise or fall in a glass tube that can be read visually. Another example is the common weather thermometer, in which contained liquid rises or falls for measuring atmospheric temperature. It is obvious that clinical or weather thermometers are not practical for measuring liquid or gas in processing because of very high temperatures and other factors.

With few exceptions, operating temperatures on process units and gas compressors are measured with thermocouples, electrical-resistance thermometers, optical pyrometers, radiation pyrometers, and specially designed instruments.

All comprehension of temperature is relative and should be stated in terms of a known, accepted scale. Temperature can be measured by

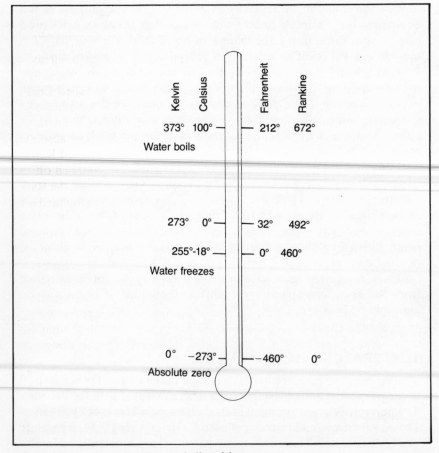

FIG. 12-1 Temperature relationships.

TABLE 12–1 Conversions

To Convert	Use Formula
Fahrenheit to C	$C = \frac{5}{9} \times (F - 32)$
	or $C = (F - 32) \div 1.8$
Celsius to F	$F = \frac{9}{5} \times C + 32$
	or $F = 1.8 \times C + 32$
Celsius to K	$K = C + 273$
Kelvin to C	$C = K - 273$
Fahrenheit to R	$R = F + 460$
Rankine to F	$F = R - 460$
Kelvin to R	$R = \frac{9}{5} \times K$
	or $R = 1.8 \times K$
Rankine to K	$R = \frac{5}{9} \times K$
	or $R = K \div 1.8$

several scales as shown in Table 12–1 and Fig. 12–1. Fahrenheit scale (°F) is commonly used in homes and industries; however, we are slowly converting to the Celsius scale (°C), which was formerly called Centigrade scale. Two additional scales in processing use the absolute zero temperature. These are called *absolute scales* because zero on these scales is the lowest temperature mankind can ever reach. The absolute scale using Celsius degrees as the temperature interval is called Kelvin (K), in which absolute zero temperature is −273.16°C (rounded off to −273°C). The absolute scale using Fahrenheit degrees is called the Rankin (R), in which absolute zero temperature is −459.69°F (rounded off to −460°F).

Heat exchange, heat balance, and many other problems involving temperature are important in all phases of both plant operations and gas compression. Means of indicating, recording, and controlling temperatures are necessary in the proper operations of equipment. Heat-affected properties of substances, such as thermal expansion, radiation, and electrical effects, are used by commercial temperature-measuring instruments. These instruments vary in their precision, depending on the property utilized and substance used as well as the design of the instrument.

Thermocouple

The thermocouple principle is based on the relation of different metals insulated from each other and joined at the ends to form a simple continuous electrical circuit (Fig. 12–2). If one of the junctions is maintained at a temperature higher than that of the other, an electromotive force (EMF) is set up that will produce a flow of current through the curcuit. The magnitude of the net EMF depends on the difference between the temperatures of the two junctions and the materials used for the conductors. The EMF is measured with a galvanometer and corresponds to temperature measurement.

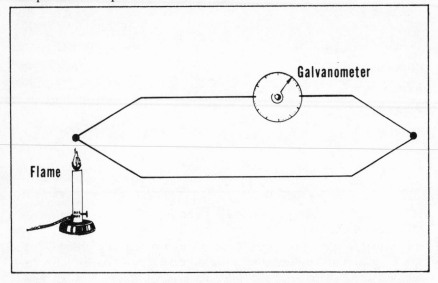

FIG. 12-2 Electromotive force is set up if temperature at one end is higher than temperature of the other end (after Babcock and Wilcox).

Thermocouples have a high degree of accuracy, are relatively inexpensive and very durable, and are versatile of application. They are convenient for recording temperature readings from one or many remote points and can be incorporated into control and regulating equipment.

Fusion Pyrometers

Fusion or change of state from solid to liquid occurs at a fixed temperature for a pure chemical element or compound, such as ice to water at above 32°F. The melting points of various materials are therefore suitable fixed points for temperature scales.

Suitable material is formed into small pyramids about 2 inches in height. These *pyrometric cones* have a known melting point at established temperatures and are sometimes used as a method of measuring very high temperatures in refractory heating furnaces They are suitable for the temperature range from 1,000 to 3,600°F. Fusion pyrometers are also made in other forms such as crayon, points, and pellets.

Vapor-Pressure Thermometer

The change of vapor pressure of a specific liquid with temperature is utilized in the vapor-pressure thermometer. The working range of a given instrument is limited to several hundred degrees and usually lies between minus 20 and plus 2,000°F.

Expansion-Type Thermometer

Most substances expand when heated, and usually the amount of expansion is almost proportional to the change in temperature. This effect is utilized in various types of thermometers using gases, solids, or liquids. Nitrogen gas-filled thermometers are suitable for a temperature range of minus 200 to plus 1,000°F.

Radiation-Type Thermometer

All solid bodies emit radiation. The amount is very small at low temperatures and large at high temperatures. For example, the temperature of hot iron can be estimated visually by its color. When iron is dark red, its temperature is about 1,000°F. and white when 2,200°F. Two types of temperature-measuring instruments are based on the radiating properties of materials: the optical pyrometer and the radiation pyrometer.

Thermometer Well

The commercial thermometer inserted in a thermometer well is widely used in the petroleum industry. It consists of a bulb filled with liquid such as mercury, pentane, toluene, alcohol or other such liquid connected to a glass stem that is properly marked to scale. The thermometer well is permanently mounted in the piping system. It is inserted in the well and packing is installed in the annular space around the stem in order to eliminate variations due to ambient temperature. These thermometers are limited to relatively small changes of temperatures.

PRESSURE MEASUREMENT

Pressure measurement is possibly the most often-made measurement in refineries and process plants. There are usually more pressure gauges used in a process plant than any other instrument. Pressure is a good, quick indication of the work done by pumps and compressors. Also, it is the most important measure of the status of operating pressure vessels.

The pressure of each vessel and tower is shown on the instrument control board of well-designed process plants. Because of its effect on boiling and condensation points and because it responds more rapidly to changes in these values, pressure measurement is sometimes used instead of temperature to monitor and control these processes. For example, heavy oils are boiled off (distilled) at a feed inlet temperature of about 700°F (371°C) in a vacuum pipe still with approximately 1 psia (pound per square inch absolute) vacuum. If the vacuum were not used, the same oil would boil off at a feed inlet temperature of about 1,100°F (593°C).

Pressure is a measure of the force exerted by a fluid due to its molecular activity. The molecules of a fluid are in continuous motion, colliding with one another and the walls of the surrounding container. This continuous impact of molecules creates a force (pressure) against the container walls.

Liquids are fluids that have definite volumes independent of the shape of the container under conditions of constant temperature and pressure. Liquids will assume the shape of the container and fill a part of it that is equal in the volume in the amount of liquid. For practical purposes, liquids are considered noncompressible because extreme pressures are required to get a very small reduction in volume.

Gases are fluids that are compressible; gases will vary in volume to fill the vessel containing them. The volume of a given mass of gas will change to fill the container. Gases cannot be contained in open-topped containers as liquids can.

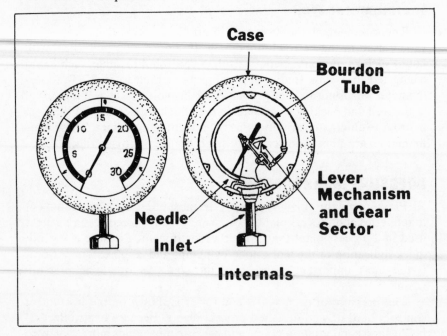

FIG. 12–3 Bourdon-tube gauge.

Bourdon Tube

The pressure gauge (or gage) is probably the earliest instrument used in refining operations. It is a direct-connected, locally mounted pressure indicator using a Bourdon tube measuring element (Fig. 12–3). The measuring element is named for Bourdon, a French engineer who invented the device in 1849. The Bourdon tube is an elastic device that functions as a pressure-measuring element because its tip movement is proportional to the pressure being measured. It is sealed at one end (the tip) and equipped with a pressure connection on the other end.

Any pressure inside the C-shaped tube exceeding the pressure on the outside will tend to straighten the tube. The movement of the sealed end of the 250° arc is used to position a pointer, a recording pen, or other indicating equipment. The amount of movement of the sealed end is formed by the tube. Usually, the movement is small and must be amplified with proper linkage if a broad scale is required. Calibrations are possible for measuring pressures up to thousands of pounds.

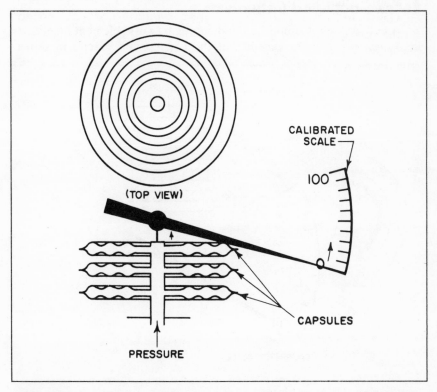

FIG. 12–4 Metallic diaphragm schematic (courtesy Ametek-U.S. Gauge).

A spiral Bourdon tube is made by winding a partially flattened metal tube into a spiral having several turns instead of a *C* shape. The spiral form produces more tip movement but does not change the operating principle of the Bourdon tube. The tip movement of a spiral equals the sum of the tip movements of all its individual C-bend arcs. Another design uses the tube wound in the form of a helix.

Diaphragms

A diaphragm element is used in measuring low pressures up to a few pounds. Like the Bourdon tube, it works because of elastic deformation characteristics. It is extremely sensitive to small pressure changes and can provide readings superior to the Bourdon tube within the range of 0 to 10 psi. It can also be used to operate an automatic control device.

There are generally two types of diaphragms used; one is made of a thin sheet metal and the other is made of nonmetallic materials. The nonmetallic diaphragms are used to measure extremely low pressures (Fig. 12–4).

Bellows-actuated Instruments

A bellows unit is similar to the diaphragm (works element on elastic deformation principle) except that the corrugations extend in series

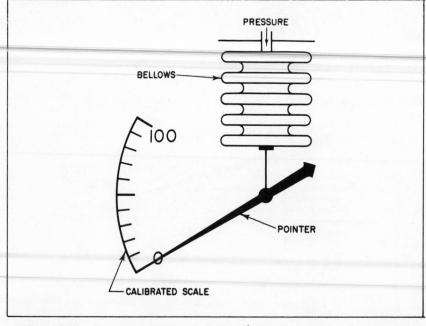

FIG. 12–5 Bellows element schematic (courtesy Ametek-U.S. Gauge).

rather than expand outward. When the pressure inside the bellows increases, the metallic discs thicken and the length of the bellows increases. This increase in length is the sum of the expansion of all the discs and is a measure of the pressure inside the bellows. Their usual range is 0 to 100 psi (Fig 12–5).

Bellows-measuring elements have a variety of uses in measuring pressures. They can be used singly or in pairs to measure absolute or differential pressure. When two bellows are used, one pressure is admitted to each bellows. By mechanically connecting the two, the difference in their expansion or contraction is a measure of the differential pressure. The most widely used bellows-actuated instrument for measuring differential pressure is the flowmeter.

U-Tube Manometers

The manometer is a fairly simple instrument used for measuring pressures from zero through a few pounds per square inch. This device is a U-shaped tube made of glass or other transparent material. One end of the manometer is connected to the pressure to be measured, and the other end is open to the atmosphere. A scale with markings is attached behind the tube, and the tube is filled about half full with fluid. The pressure will force the fluid to move in the tube until the weight of the displaced fluid is equal to the force of the exerting pressure. The difference between the height of the fluid in the two sides of the U-tube is the pressure measurement. Because of their fragile construction and limited range, they are seldom used except where another instrument is impractical.

Electrical Pressure-measuring Instruments

Elastic pressure-measuring elements convert the volumetric changes caused by pressure into mechanical movement. The most common instrument of this type is the *strain gauge*. Basically, the strain gauge consists of a small wire arranged so a change in the pressure stretches the wire. Its electrical resistance is changed as the wire stretches, and this change can be measured and converted to pressure change.

LIQUID LEVEL MEASUREMENT

The sight glass is used in certain tanks and vessels for a direct reading of liquid level. It consists basically of a transparent column, usually glass, attached to the tank or vessel by suitable leakproof fittings. There is usually a scale attached to the column by which the liquid level can be observed when the valve(s) is opened. If the vessel is at atmospheric pressure, the top end of the column can be vented.

Float-type Units

Liquid level by float-type units is determined from the buoyancy of a floating body partially immersed in the liquid. Buoyancy is the upward force exerted on the floating body by the liquid.

The float is usually a hollow metal ball that is mechanically connected to the measuring instrument. Suitable leakproof connections are installed between the float and instrument when necessary.

Liquid level has no absolute value and is always relative to a reference point such as the tangent line on a vessel and the bottom of the vessel. Liquid level is easily understood and can be directly observed.

FLOW CONTROL

One of the important factors that determines the physical setup of a refinery or gas plant is the rate of fluid flow. After determining flow rate, the next problem is providing the equipment that is best suited to do the job. Some control functions can be accomplished manually; however, all control functions can be accomplished better automatically. Automatic control systems should have a measuring element, controlling element, and a final control valve.

The *measuring element* could be a Bourdon tube, bellows-actuated instrument, float valve, or suitable elastic pressure-measuring element. It must detect any deviation in the controlled medium variable.

The *controlling element* produces a signal when any variable occurs. The signal can be produced by various pieces of equipment. Some employ an electrical sensor used with a conventional measuring element. These sensors convert the force or movement of the element to a measurable characteristic of an electrical circuit. Movement of the element can shift a contact across a resistance coil. It can also change the position of a coil in a magnetic field or increase or decrease the gap between the conductors in a capacitor. Other controlling elements can utilize pneumatic sensors.

Final control is often accomplished with a diaphragm motor valve. It consists essentially of a diaphragm, valve stem, loading spring, valve body, packing, seals, inner valve, and valve seat. Different arrangements can be used to operate in normally open or normally closed diaphragm motor valves (Fig 12–6). A control valve must convert the control signal from the controlling element into a controlled rate of flow. Control valves may require certain accessories such as positioners and volume boosters to function properly.

GAS REGULATORS

There are several varieties of mechanical devices designed for controlled reduction of gas pressure. Any gas regulator matches the flow of gas through the regulator to the demand for gas downstream, and it

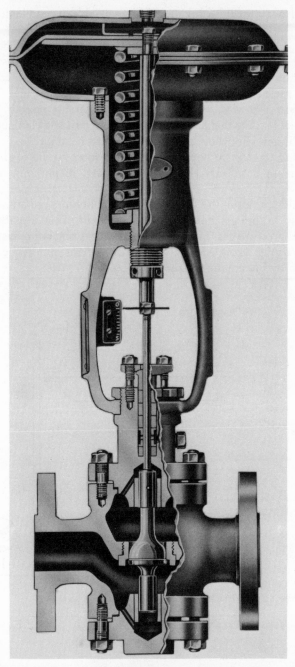

FIG. 12–6   Single-ported control valve (courtesy Masoneilan).

should maintain the system pressure within certain limits. All gas regulators have three basic components in common regardless of size, type of service, or complexity: a regulating element, a sensing element, and a loading element.

The *regulating element* creates a restriction in the gas line that is decreased or increased in size to vary the gas flow rate. It is usually a valve, although it can be an expandable sleeve or any device that will modulate the flow.

The *sensing element* responds to the downstream pressure and controls a variable force that acts to position the restricting element. It could be a diaphragm, Bourdon tube, bellows, piston, or gauge.

The *loading element* applies the needed force to the restricting element to cause it to vary. It can be a spring, diaphragm, weight, or piston.

Several varieties of the spring-loaded diaphragm regulators are commonly used in fuel gas lines to field equipment. They are also used to supply low-pressure gas for homes. A typical gas regulating system could be similar to Fig. 12–7. More complicated systems remotely control the diaphragm by pneumatic instrumentation.

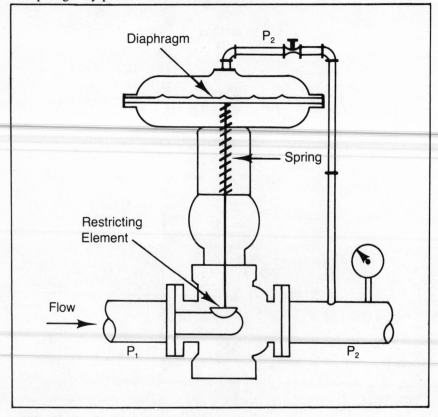

FIG. 12–7 Typical gas regulator.

GAS MEASUREMENT

Natural gas is a clean, efficient source of energy and must be conserved whenever possible. This means using efficient handling facilities and good measurement practices. The volume of gas is measured to determine how much is being consumed or sold.

Gas volumes can be measured with orifice meters, positive-displacement meters, turbine meters, venturi meters, rotameters, and other specially designed equipment. One common type familiar to most people is the typical domestic meter used by gas utility companies.

Measurement of gas requires that the unit of volume be defined. This means that the unit, pressure base, temperature base, and specific gravity must be determined or specified. Usually a cubic foot of gas is the amount required to fill a cubic foot of space at 60°F and 14.73 psia. Gas billing is normally priced on 1,000 cubic feet consumed.

CHAPTER 13

Marketing

Marketing, the delivery of the finished product to the customer for value received, is the final step in the process that begins with the search for oil. Over the years, two methods of selling petroleum products to the consumer have developed. The first is by an *integrated company*. That is a company large enough to engage in all aspects of the oil industry—one company (or its affiliates) that prospects, drills, produces, refines, transports, distributes, and sells at retail at a service station bearing the company emblem.

The other sales effort is undertaken by *independent marketers* who buy either the overstock of name-brand refiners or buy from independent refiners and distribute through their own system of transport, bulk plant operators, and service-station owners. Each system has its own particular merits.

DISTRIBUTION

Consumer products such as gasoline and home heating oil may be sent from the refinery via pipeline, tanker, barge, or railroad to a terminal located near a large population center and from there be either sent on to a bulk plant serving a smaller city or directly to service stations by tank truck. In the case of heating fuel, the jobber or bulk plant operator will make home deliveries using smaller trucks.

INTEGRATED COMPANIES

During the first half of the century, those oil companies involved in all phases of the industry set up their own distributing and sales organizations. As roads improved and the number of automobiles increased, thousands of service stations were built and great sums were spent on advertising and on development of trademarks and brand names. Generally, the *jobber* was an independent businessman who contracted with a particular company as their agent to sell their brand of product at wholesale in a given trade area.

The filling-station operators fell into two categories: operators of company-owned stations and owners of their own buildings and locations. In either case, owning a filling station came to occupy an important place in American society. For those who operated company-owned locations, it presented the opportunity for a person to go into business for himself and become his own boss for a very modest investment—one that was usually within the reach of most people. A person could own a filling station by purchasing his original inventory, paying a rent based on the number of gallons pumped, and agreeing to operate the station in accordance with company rules.

Great amounts were spent to ensure that stations were built to look alike so that a customer could easily recognize his favorite brand no matter what part of the country he was in, to maintain cleanliness, and to train employees in courtesy and salesmanship.

By mid-century a driver knew that he could pull into a company station anywhere, use a clean restroom, have his tank filled with a quality product he trusted, and at the same time have his oil, water, battery, and tire pressure checked, his windows washed, and his car swept out. If he needed a battery, tire, wiper blade, or lubricant, he could purchase these items bearing the same brand name as the gasoline at the pump and have the work done as he waited.

Those operators who owned their own locations generally offered these services, but many offered petroleum products for sale as a sideline, such as a country grocer who would have a pump or two out front and did little or no service or repair work. Despite the number of both types of stations, there was still room for independents.

INDEPENDENT MARKETERS

The need for independent marketers grew from a few simple economic facts. First, petroleum is and is not a seasonal product. It is distinctly different from a farm crop that matures at a given time, is harvested, stored, and then sold at a more-or-less even rate of demand all year.

Petroleum can be produced all year, but the demand fluctuates seasonally. For example, heating oil is needed only in the winter; more gasoline is sold during summer vacation months and on holiday weekends. If all the refineries operated at capacity year round, there would be an excess of gasoline that company stations could not possibly sell. Storage is impractical because of the cost of holding such a bulky product and also because gasoline does not store well. Over a period of time, it will deteriorate in quality. Also, understandably, the major oil companies were interested in building stations in areas of high traffic volume. This left many potentially profitable sites available in areas the majors did not wish to serve. Out of these two factors, the market for the independents grew.

If the big companies built too few refineries, they could have found themselves running short of product at critical times. If they built too many, they would have an excess of product on their hands. And there were thousands of customers who lived outside the major's trading areas, or who were more concerned with price than with brand names.

The independent marketers stepped into the picture and helped alleviate the refiner's problem by buying up all of the excess or overstock produced. Some of the independents, such as J. B. Saunders Jr.'s Triangle refineries, grew so large they were eventually contracting for the entire output of certain refineries. They in turn either resold to small, independent jobbers who supplied a few filling stations in their community or they set up their own system of distribution and retail outlets. Sometimes this consisted of buying station property from the majors or building their own chains of stations. It was not unusual for an independent to sell to the public under 20 or 30 different brand names.

In general, these stations were small, no-frills affairs that offered the customer gasoline a few pennies a gallon cheaper than the majors and sold motor oil in bulk rather than in one-quart cans. Saunders, whose operation began with a one-man office opened during the middle of the Great Depression, eventually became the largest independent marketer in the world with his own fleet of railway tank cars, pipelines, barge lines, and terminals.

CREDIT

Long before World War II, the major companies learned they could increase their retail sales and get repeat business by offering certain customers the privilege of charging their gasoline purchases on a monthly basis. After the war as consumer credit grew, so did the use of gasoline credit cards. The number of potential credit customers expanded. To meet the competition, the companies began offering extended terms for tire and accessory purchases.

Credit privileges were extended into other areas; some companies affiliated with motel and restaurant chains and customers could charge an entire vacation on their gasoline credit card. To attract new customers, mass mailings of credit applications were made to certain groups, such as graduating college seniors.

OTHER RETAILING ACTIVITIES

Everyone is familiar with the traditional advertising methods of the majors such as newspaper and magazine ads and the sponsorship of radio and television programs. Other direct customer appeals included giving away free road maps and premiums at service stations and the use of trading stamps. Some companies offered free trip routing and planning services, and at least one went into the travel-club business, offering not only trip planning but insurance, bail bond protection, and towing service.

Some companies were quick to spot the opportunities offered by aviation and sponsored early fliers and endurance flights. Not only did this bring the company's name before the public; it also gave them credence with the fledgling aviation industry, since the connection allowed for research and development of flight-oriented products as well as advertising. Other companies made similar arrangements with boating and auto racing enthusiasts.

REGULATIONS

Aside from alcohol and tobacco, few other consumer products are as carefully watched over as motor fuel. Each gallon of gasoline is taxed by both federal and state governments (fuel sold for strictly off-the-road equipment, such as farm machinery, does not pay the same tax). States guard this source of income so jealousy that truckers have to post a fuel bond to drive through a state even though they may not happen to purchase any gasoline or diesel fuel in that state on that particular trip. Such fuel taxes have become an important source of revenue used for highway construction, school support and other areas of public financing.

Service stations must meet the fire and safety regulations set forth by the state fire marshal's office as well as comply with any applicable OSHA rules. In addition, there is usually a state agency that periodically inspects filling stations to make sure that their pumps are accurate and give the customer full measure. The quality of the product may also be checked to make sure it does not contain water or other contaminants and meets minimum octane requirements.

MARKETING CHANGES

Two developments in the third quarter of the century brought about some major changes in the marketing of gasoline. The first was the requirement by the government that new cars must be manufactured to use only unleaded fuel. This meant that refiners had to manufacture vast amounts of unleaded gas and that service stations had to replace existing storage tanks or install new tanks and pumps to handle the new fuel. Since tanks are buried underground, often beneath the station's apron or driveway, this amounted to a major expense.

Then increased demand and decreased domestic supplies brought more imported oil from abroad, most of it from OPEC (Organization of

Petroleum Exporting Countries). Well aware of the situation the industrialized nations were in, these countries dramatically increased the price of crude oil.

This increased the companies' cost of business, so measures were taken to cut down on expenses. These included closing down unprofitable or marginal stations and doing away with give-aways such as free maps and trading stamps. There was also a reduction in the solicitation of credit-card accounts. Some companies reduced their accounting expenses by accepting bank charge cards.

Since the retail price of gasoline jumped accordingly, many stations offered their customers the option of self-service. For a reduction in price of a few cents per gallon, the customer learned to fill his own tank, check his own oil, and clean his own windshield. Those who wished full service as they had received in the past paid full price for their gas.

Also gone from the American scene were the price wars, during which companies would slash pump prices below a competitor's to attract customers. And the difference in retail price between name-brand gasoline and that of independent stations shrank. At the same time, consumers were being warned by both the companies and government agencies to expect even higher gasoline prices in the future.

CHAPTER 14

Tomorrow's Energy

As we step into the '80s, the phrase *energy crisis* is becoming a household word. Although petroleum forms all the time as the normal sedimentation process goes on, the time span for renewal is millions of years. We're depleting the available reserves in a few decades. Inevitably, alternate sources of energy will have to be explored as we move into the 21st century.

OIL AND GAS

Most government and industry studies agree that oil and natural gas, which now provide almost three-fourths of all U.S. energy, will still provide between 55 and 65% of our energy needs between 1985 and 2000. There is no way of predicting with any degree of reliability just how much petroleum there is to be found worldwide, but experts speculate there may be as much as two trillion barrels. This means there are still sufficient resources of petroleum to provide an orderly transition from petroleum to alternate-energy sources during the next 50 years (barring unexpected interruptions in imports).

Domestic Potential

The U.S. has a tremendous resource potential for both the discovery of new oil and gas reserves and the recovery of additional reserves from known reservoirs. With improved technology, prices keeping up with costs, and the right economic and political climate, the U.S. has an additional 82 billion barrels of undiscovered natural gas. That could mean a 33-year supply of oil and a 36-year supply of natural gas at current levels of production.

Several items now block some avenues for production and exploration. One is the influence of prices. With the decontrol of oil in 1981, oil companies could finally receive the revenue they needed to fund experiments in enhanced oil recovery. Natural gas, however, is as yet still controlled. Exploration cannot be expected to climb until prices for produced gas are allowed to rise without restriction. Another possible restriction is access to public lands. An estimated two-thirds of the potential oil and gas discoveries may lie on public lands onshore and offshore. Yet many of these areas have never been made available for exploration and development. A review of these policies as well as environmental laws may be needed to increase production of reserves.

Today our production in barrels of oil equivalent which includes natural gas and natural-gas liquids is about 19.5 MMb/d. These will decline in production to about 8 MMb/d by 1990 and less than 2 MMb/d in 2000. Obviously, an all-out effort at increasing oil and gas production is necessary if we are to have any hope of stabilizing existing production declines. This can be done by exploring new areas, increasing production levels on existing mature areas, and implementing enhanced oil recovery methods to recover as much oil as possible.

HEAVY OIL

Heavy oil is a very dense, highly viscous crude oil that is similar to ordinary petroleum. However, it is so gummy that it must be heated or otherwise coaxed to flow before it can be pumped out of the ground. On the scale of petroleum products, it lies somewhere between the heavier grades of ordinary crude and tar sands. For comparison, very light crude is rated at 34. Heavy oil is rated at 20. Recoverable reserves in the U.S. are estimated somewhere between 10 and 15 billion barrels. Total reserves worldwide are estimated to be more than 10 times that amount.

The most common method of producing heavy oil is to pump steam into the ground through injection wells. The steam thins the oil and drives it out a production well. When refined, heavy oil tends to yield more residual (heavy) fuel oil and less gasoline and other light products. As a result, expensive upgrading equipment is needed to convert some of the heavy products into the more desirable ones.

The major stumbling blocks have been economic and environmental. In California where most of the nation's heavy-oil reserves lie, deep heavy crude can cost as much as $10/barrel to produce, then $3–5 per barrel to refine. Also in California, environmental regulations are the most stringent. This creates a deterrent to production.

COAL

Coal is available in abundance. The U.S. has known recoverable reserves of 220 billion tons of coal—eight times the nation's proven oil reserves and enough to last more than 300 years at the present rate of production. In fact, the U.S. has the largest recoverable reserves in the world.

In the long term, coal could become a major source of synthetic fuels and gaseous fuels. In the short term, coal could be burned directly in some electrical utility and industrial boilers that now use oil and natural gas. If 20–30 plants were converted to coal, more than 200,000 barrels per day would be saved.

Essentially, two methods of mining are used to extract coal. *Deep mining* is used when coal deposits are not accessible from above. This method is labor intensive and hazardous. *Surface mining* involves stripping away 30–225 feet of topsoil and rock (overburden) and mining coal deposits directly below. After the coal is mined, the overburden is replaced and replanted. This method is safer and more economical, but it has been criticized for its ecological effects.

For the most part, the impediments to coal development have been related to the environment. Strip mining is opposed by many. Coal slurry pipelines are efficient and economical but use large amounts of water—particularly in the arid West. Burning coal gives off noxious fumes that pollute the air. Consequently, some fear that increased use of coal will lead to more air pollution.

Coal is a plentiful and relatively inexpensive resource. The U.S. has adequate reserves to supply a large portion of this nation's energy needs for decades. The technology for production is already developed and the cost is relatively low. However, environmental costs will be high. Easing environmental restrictions would probably do more to encourage coal development than any other type of incentive. Modifying the Clean Air Act and relaxing strip-mining laws could help. Nevertheless, the future of coal remains an ultimate showdown between environmental goals and energy needs.

GASOHOL

This fuel, a blend of 10% anhydrous ethyl alcohol (ethanol) and 90% gasoline, is one means of stretching the available supply of fuel. Ethanol is produced by distilling grain alcohol from virtually any agricultural resource, e.g., corn, sugar cane, wheat, potatoes, or even fermented garbage (biomass). However, the ethanol made from grain is more expensive to produce than the gasoline it replaces. The process also consumes more energy than it produces. And in a world where children are starving, a moral question must be asked whether or not we are right burning surplus grains in our gasoline tanks.

One alternative to this process would be to substitute methanol for the ethanol. Methanol is an alcohol made by gasifying coal. It can be produced by a proven indirect process where the coal is first gasified and then converted to methanol.

Under present initiatives, the production of ethanol for gasohol is likely to be limited. The complex distillation process makes ethanol expensive. The time and cost required to build new capacity will limit the expansion of the resource for the next few years. Methanol derived from biomass and coal is expected to play a larger role in the long term, but only if new technology is developed.

The most optimistic projections for alcohol fuel use indicate that about 40,000 barrels of imported crude per day could be displaced by 1985, largely by supplementing with ethanol. However, this would greatly harm U.S. food needs. Methanol could be used, largely in turbines and other stationary equipment. However, it will probably be 1990 before 400,000 barrels per day will be produced. And regardless of ethanol or methanol, the price will be double that of petroleum products.

SHALE OIL

Oil shale is a fine-grained rock that contains varying amounts of a solid organic material called *kerogen*. When heated to nearly 900°F the kerogen decomposes into hydrocarbons and carbonaceous residue. The cooled hydrocarbons condense into a liquid called *shale oil*. This can be refined into a useable fuel.

Almost two trillion barrels of oil are trapped in shale formations in a 16,000-square-mile area that extends into Colorado, Utah, and Wyoming. While these oil shale deposits are potentially large, only 600 billion barrels (one-third) are recoverable.

Presently, there are two basic processes for extracting the oil. In one, the shale is mined and heated on the surface until the oil comes out. In the other, a well is dug, explosives are set, and a huge underground cavern of rubble results. The shale is then heated underground and the oil is pumped from the bottom of the cavern. This is expensive, but there is no rock rubble left behind.

Several technical, environmental, and regulatory obstacles must be resolved. First, the methods are still in the text stages. No one yet knows the full capabilities of the project. Second, water is necessary for the process. In the arid West where oil shales abound, water is scarce. Finally, approval from state and federal agencies is always delayed. Paperwork can hold up a project for months and even years.

Of the various new technologies to produce synthetic fuels, shale oil is among those closest to commercialization. Based on a 50% chance of technical success and commercial implementation, the proper incentives, and cooperation from agencies, the U.S. could produce 360,000 barrels per day of shale oil by 1990. However, the price tag will be high.

COAL LIQUEFACTION AND GASIFICATION

There are several technologies for producing gas from coal that are ready for commercial use. New technologies for producing liquids and gas from coal will be ready for commercialization during the next few years. These fuels could provide energy independence from imports.

Through either a liquefaction or a gasification process, the coal releases liquid or gas. This material can be transported through existing pipelines and can be adapted in refineries. Best of all, the liquid and gas can be used for cars and other forms of transport—one of our major energy concerns.

The key to the process is hydrogen-enriched coal. Ordinary coal has 16 atoms of carbon for every atom of hydrogen. Natural gas, on the other hand, has 4 atoms of carbon for every atom of hydrogen. Under high temperatures and pressure, coal's chemical structure can be broken down to admit hydrogen atoms to its molecular structure. This way, a hydrocarbon fuel and gas are formed.

A commercial-sized plant could produce 58,000 barrels of synthetic crude per day using 40,000 tons of coal per day. This would be a size comparable to the SASOL plant in South Africa. If the direct liquefaction process were used, 90–110,000 b/d could be produced using 40,000 tons of coal. As with shale oil, however, environmental and regulatory policies will greatly affect the outcome and economic feasibility of the program.

Energy companies are prepared to build modular-sized plants for producing gas and oil from coal. These will provide information on the environmental obstacles as well as economic uncertainties that need to be overcome. Since these are all based on using the United States' most abundant domestic energy source, they offer the opportunity to expand domestic energy production and provide a resource base on which to build a major domestic energy industry.

TAR SANDS

The U.S. has some tar sands—mainly in Utah—but the largest deposits are in Canada where bituminous sands cover 12,000 square miles and contain 900 billion barrels of oil. Tar-sand deposits in the U.S. are estimated at 130 to 200 billion barrels. Once extracted, the tar-like sands are mixed with hot water and steam to form a slurry. Then a process breaks the material down into sand and bitumen, which is then cleaned, cracked, and upgraded with hydrogen and a catalyst.

Producing oil from tar sands is similar to the processes used for oil shales. It can either be strip mined and heated or heated in situ and pumped.

The development of oil from U.S. tar sands is somewhat constrained by environmental laws and regulatory delays, but the major obstacle is technological. Most U.S. deposits are relatively small, and in-situ technology is still relatively new. Until a method can be found to extract the oil more cheaply, developers will probably opt for one of the other alternatives, which is more economical.

Thus, the probabilities are quite low that U.S. tar-sand deposits can be developed in the next 7–10-year period on a commercial scale. This is based on the limited extent of the deposits, the lack of sufficient surface mineable deposits, and the new phase of in-situ recovery.

OTHER METHODS OF ALTERNATIVE ENERGY

We all hear about advances in solar, nuclear, wind, thermal, and hydrogen technology. At the present time, however, none of these alternatives can provide ample amounts of energy needed to fuel our cars or heat our homes. They provide at best a way of cutting down our consumption of petroleum-based fuels. In the future, these sources will be investigated. In the immediate forecast, however, we will need to maintain a smoother transition, using the refineries that are at our disposal and the vast wealth of petroleum knowledge in extracting materials from the earth.

Glossary

abrasive drilling The use of a harder mineral or substance than the rock being drilled through as a drilling medium.

acidizing The introduction of acid into a well to dissolve deposits of alkali material to open passageways for the oil to flow through.

acoustic log A measuring device that uses sonic waves to directly measure lithography and porosity. Also used to detect poor cement bonding to casing.

air balance beam pumping unit Device using compressed air to balance the weight of the beam.

air bursts A marine geophysical technique in which bursts of compressed air from an air gun towed by the seismographic vessel are used to produce sound waves formerly generated by high explosives.

air drilling Rotary drilling system using compressed air instead of fluid as the circulation medium.

air injection A secondary or tertiary means of recovery in which compressed air is introduced into the formation to force the oil out.

alkanes This series, derived from petroleum, has its carbon atoms arranged in a straight chain. It includes methane, ethane, propane, butane, pentane, hexane, and heptadecane.

alkylate Gasoline grade components manufactured from refinery gases.

alkylation The reaction of alkenes or olefins with a branched chain alkane to form a branched, paraffinic hydrocarbon with high antiknock qualities.

all-levels sample Taken by submerging a stoppered beaker to a point as near as possible to the draw-off point, opening it, and raising it.

allowable The amount of oil or gas a well is permitted to produce under the orders of a regulatory body.

alluvial Pertaining to the sediment deposited by water flow.

alpha I A new cable device that uses less energy to operate than a conventional pumping unit.

angular unconformity Strata lying at an angle across the folded and tilted edges of the beds below it.

anticlines Arches or upfolded rock formations.

API gravity The standard method of expressing the gravity, or unit weight, of petroleum products.

arch A rock formation that folds upward like an inverted trough.

aromatics Cyclic hydrocarbons, originally found in aromatic gums or oils.

artificial drives Drive other than natural means—in-situ combustion, waterflooding, etc.

artificial lifts Any mechanism other than natural reservoir pressure of sufficient force to make the oil flow to the surface.

average sample Averaging of two or more samples.

automatic tank batteries Lease tank batteries equipped with automatic measuring, gauging, and recording devices.

B

barefoot Well completed without casing in a sandstone or limestone formation that gives no indication of caving in.

barite A mineral often used as one of the components of drilling mud to add weight or body; barium sulphate.

barrel 42 U.S. gallons.

batch A shipment of a particular product through a pipeline.

batch interface The point where two shipments of product touch in a pipeline, e.g., a batch of gasoline followed closely by a batch of kerosene.

batch separator A device used to keep shipments apart in a pipeline so that different liquids do not intermingle.

batching sequences The order in which product shipments are sent through a pipeline.

b/d Barrels per day.

bead Deposit of molten filler metal laid down during the welding process.

benzine Light petroleum distillates in the gasoline range. Still used as a synonym for gasoline in some European countries.

Big Inch The 22-inch pipeline built from Texas to the East Coast by the government during World War II.

bits The cutting tools used on the working end of the drill string.

blowout When excessive well pressure runs wild and blows the string and tools out of the hole.

blowout preventer Device consisting of a series of hydraulically controlled rams that can be triggered instantly to seal off a well.

bottom drive Pressure from salt water under the oil, forcing it upward.

bottom fraction The heaviest components of petroleum—those at the bottom of the barrel after the lighter ends have been removed.

bottom sample Obtained from the material at the lowest point in the tank.

bottom water Water located at the bottom of a reservoir under the petroleum accumulation.

bright spots White areas on seismographic recording strips that may indicate the presence of hydrocarbons.

BS basic sediment.

BS&W Basic sediment and water.

Burton process Early process developed by Dr. William C. Burton that increased production of light products by using heat and pressure.

C

cable-operated long stroke Pump that uses a cable from a tower instead of a walking beam to lift the sucker rods.

Cambrian Geologic period from about 600,000,000 B.C. to 500,000,000 B.C.

cannel coal Bituminous coal that burns with a heavy smoke.

cap bead Final welding process on pipeline joints.

capillaries The minute openings between rock particles through which oil and water are drawn.

capillary action The upward and outward movement of oil and water through the pore spaces in rock.

carbon A natural element found in hydrocarbons.

carbon black A "soot" produced from natural gas used in the manufacture of tires, etc.

carbon-dioxide injection A secondary or tertiary means of recovery that uses compressed carbon dioxide to force the oil to the well.

casing Pipe used in a well to seal the borehole to prevent fluid escape and to keep the walls from collapsing.

casinghead gasoline Natural gasoline—actually the condensate from natural gas.

catalysis A process in which the chemical reaction rate is affected by introduction of another substance.

catalyst A substance that is used to slow or advance the rate of a chemical reaction without being affected itself.

catalytic (cat) cracking The use of catalysis to break petroleum down into its various components.

cat head Spool-shaped hub on a winch shaft connected to the draw works around which a rope may be snubbed.

caustic injection Introducing substances into the formation to increase porosity by breaking down particulate matter and also using pressure to increase the flow of oil.

caverns Larger openings between the rocks in a formation.

cellar Area dug out beneath the drilling platform to allow room for installation of the blowout preventer.

cement Mixture which is used to set the casing firmly in the bore hole. A slurry, it is allowed to set until it hardens.

cementation The natural filling in of the pore spaces in a reservoir by limestone.

cementing Pumping the cement slurry down the well and back up between the casing and the bore-hole. Once hardened, the cement is then drilled out of the casing.

Cenozoic Geologic era from about 63,000,000 B.C. to the present.

centralizers Devices fitted around the outside of the casing as it is put in place to keep it centered in the hole.

channelization The act of taking a shallow, winding waterway and straightening and deepening it so it can be used for transportation.

chemical precipitates Material formed in place under the earth's surface by the action of dissolved salts.

cheese box Early still using a vessel that resembled a cheese container.

Christmas tree Array of valves, pipes, and fittings placed atop a free-flowing well.

circulating system That portion of the rotary drilling system which circulates the drilling fluids or mud.

clamshells Hinged, jaw-like digging tools used on a dragline.

clean circulation In rotary drilling, when the drilling fluid returns clean without cuttings coming to the surface. A method of refining that eliminates much of the residuum.

clearance sample A sample taken from 4 inches below the level of the tank outlet.

combination trap A reservoir formed by folding, faulting, and porosity changes.

common carrier A public, for-hire transport, i.e., bus, truck, train, pipeline, airline, etc., that falls under the jurisdiction of the Department of Transportation.

combination drive Two or more natural mechanisms present in a reservoir such as water or gas-cap drives.

completion Finishing a well. Preparing a newly drilled well ready for production.

composite sample Composed of equal portions of two or more spot samples.

composite spot sample A blend of spot samples mixed in equal portions.

compressor Mechanical device used in the handling of gases much as a pump is used to increase the pressure of fluids. Also used to increase air pressure.

compressor station Placed at selected intervals along a gas pipeline, these units maintain the pressure necessary to keep the gas flowing through the lines.

conceptual models By using various types of information available, a geologist can illustrate on paper what underground structures probably look like.

Condeep Offshore drilling and production platform designed for use in the North Sea.

conductor Outer pipe near the top of the well used to seal off unstable formations or to protect ground water near the surface.

connate water Salt water not displaced from the pore spaces which coats the surfaces of the larger openings and fills the smaller pores.

continental environment Sediments deposited on land by the wind.

contour maps A map on which the elevation in height (or depth) is visually indicated.

core Literally a plug lifted or cut out of the earth at a predetermined depth.

core drilling Using a special bit for the purpose of cutting a core.

core sampling Taking out a core for geological examination of the composition of the strata at a particular depth.

coring bit A hollow bit designed to make a circular cut for a core sample.

correlation markers Indicators used on a map to cross-reference one particular feature to another.

corrosion A complex chemical, physical, or electrochemical action destructive to metal.

cracking Process of breaking crude oil down into its various components.

cradling Lifting of the welded and wrapped pipeline into the trench.

cross process Cracking process of the 1915–1920 era developed by Gasoline Products Co.

cross-sectional map A vertical slice map illustrating features above and below ground.

crown block Pulley at the top of the derrick that raises and lowers the drillstring. Attached by lines to the traveling block and hook.

crude Oil, unprocessed, just as it comes from the formation.

crust Outer covering of the earth.

cut   A particular hydrocarbon fraction.

D

daily drilling report   Completed every morning by the tool pusher from the records kept of the activities of the drillers from the three previous tours.

darcies (d.)   Unit of measurement of permeability. Named after its originator, Henry D'Arcy.

day-work basis   Contractual arrangement of payment for drilling where the drilling contractor is paid by the day instead of by the foot.

deltaic plain   Depositional deposit area of sedimentation beneath the alluvial plain but above the normal marine.

decompression chamber   A sealed room in which the atmospheric pressure may be varied. Used to isolate divers and gradually adjust their bodies to differences in pressure to prevent or treat the bends.

Department of Energy   Federal umbrella agency overseeing all aspects of energy production.

depletion drives   A situation where the oil does not come in contact with water-bearing permeable sands, thus must depend on either solution-gas or gas-cap drive as a lifting mechanism.

derrickman   Member of a drilling crew who works on the tubing board of the rig and handles the pipe joints.

detritus   Fragments of minerals, rocks, and shells moved into place by erosion.

deviation   Directional change from the absolute vertical in drilling.

Devonian   Geologic period from about 405,000,000 B.C. to 345,000,000 B.C.

direct detection   Method of reading the possible underground location of hydro-carbons from white areas on seismographic record strips.

disconformity   Situation where the layers above and below the unconformity are parallel.

distributing lines   Lines running from the product line to the various markets.

diverter   A bypass system used in drilling at sea that vents natural gas away from the drillship in the event of a blowout.

DOE   *See* Department of Energy.

dolomite   A sedimentary rock. Possibly formed from limestone through the replacement of some of the calcium by magnesium.

dolomitization   Ground-water action causing limestone to change to dolomite. As it does so it shrinks, making larger openings.

domes   An upthrust in the earth's surface caused by the forcing upward of salt or serpentine rock by pressure from below.

dope gang   That portion of a pipeline crew assigned to covering the line with a protective coating.

DOT   Department of Transportation.

double   Two joints of pipe fitted together.

drain sample   A sample obtained at the discharge valve.

draw works   The hoisting equipment of a drilling rig.

dredge   Vessel designed to scoop or pump out a trench or channel in a river.

driller The man in charge of the drilling crew on each tour.

drilling mud A fluid consisting of water or oil, clays, chemicals, and weighting materials used to lubricate the bit and flush cuttings out of the hole.

drilling program The planning process for assembling all the personnel, equipment, and supplies for drilling and completing a well.

drilling template Mechanical device placed on the ocean floor with orifices through which the drillstring passes and to which the marine riser is attached.

drillship Large vessel designed for offshore drilling operations.

drip Wellhead device for tapping off natural gasoline.

drives The energy force present in a reservoir that causes the oil to rise toward the surface.

Dubbs process The clean-circulation process of cracking developed by C. P. Dubbs that greatly reduced the amount of coke deposits.

dry hole A duster; a well that fails to hit oil or gas.

dry-hole money Money paid to an operator by a lease owner or owner of a lease near the well site if the well fails to strike pay sands. Even though no oil is found, valuable information about the land is gained.

DWT Dead-weight tons.

dynamic positioning Means of keeping a drillship positioned exactly above the drillsite by transmitting position signals from the ocean floor to the ships' thrusters.

E

edge water Water around the edges of a reservoir that presses inward.

Ekofisk North Sea offshore drilling and production complex.

electric log Used to determine porosity and lithography.

electrodrills Rotary drills powered by electricity.

eminent domain Legal concept which gives government, and those so authorized by government, superceding access to land or property.

Energy Regulatory Administration Federal agency that sets energy policy such as pricing.

Energy Resources Development Agency Federal agency charged with promoting and developing sources of energy. Now part of the Federal Energy Administration.

environmental impact study Detailed report required by the Environmental Protection Agency before any major construction project can be begun to determine its impact on the surrounding environment.

Environmental Protection Agency Federal agency charged with protection of the environment—air, water, etc.

environments Deposits of sedimentary rock. *See* continental, transitional, and marine.

erosion The process of wearing away by water, ice, wind, or wave action.

escape capsule Floatable, watertight escape vessel used at offshore rigs for emergency evacuation.

explosive fracturing Use of explosive charges to shatter a formation. May be fired through the sidewalls of the well.

F

fault A fracture in the earth where the rock on one side moves.

Federal Energy Regulatory Commission Government agency that regulates energy in interstate commerce (FERC).

filler bead Placed over the hot passes to continue building up the weld.

firing line The part of the pipeline crew that makes the finishing welds.

flares Devices that burn off excess natural gas at a well or production site.

flow Movement of petroleum through the reservoir.

folds Buckling of the earth's strata caused by movement. *See* upfolds.

fourble Four joints of pipe fastened together.

fraction Each of the separate components of crude oil or a product of refining or distilling.

fractionating columns Tall metal column used in processing liquid petroleum into its various components.

fracturing Artificially opening up a formation to increase permeability and the flow of oil to the bottom of a well.

Frasch process Process developed by Herman Frasch using cupric acid to treat sulfur-bearing crude oil.

free gas Gas occurring in a reservoir that is separate from the oil.

G

gamma-ray log A nuclear log that measures natural radioactivity to determine lithography.

gas Natural gas.

gas cap Gas trapped above the oil in a reservoir.

gas-cap drive If a gas cap located above the oil is tapped, the gas continues to expand and force the oil downward to the bottom of the well and then back up the bottom of the well and then back up the bore.

gas drilling Drilling process using gas as the circulating system, similar to air drilling but using natural gas.

gas lifts Inducing gas into the reservoir to force the oil out.

gas pipelines Pipeline designed to transport gases such as natural gas.

gathering lines Lines from lease tank batteries to the crude trunk line running to the refinery.

gauge ticket Written record kept by a gauger or pumper indicating the amount and quality of production.

gauger Person who measures the amount of oil in a lease storage tank or the amount of oil entering a pipeline.

geophones Microphones plugged into the earth's surface used to detect seismic waves.

go-devil Cleaning device sent through a pipeline (*see* pig).

graben Valleys between high peaks of igneous rocks that become filled with sediment.

gravity meter An instrument that indicates the density of rock formations, measures the gravitational pull of buried rocks, and provides information about their depth and nature.

gun barrel tank   A settling tank placed between the pumping unit and other tanks, normally fitted with a connection at the top to separate the gas. Usually smaller in diameter and taller than the other tanks.

H

heavy crude   Thick, sticky crude oil of heavy specific gravity.

holiday   Gap left in the protective coating of a pipeline.

holiday detector   Electronic device used to detect gaps in pipeline coating.

Holmes-Manley process   Texas Co. cracking process of 1915–1920.

horsehead   End of a pumping beam to which the polished rod is attached; resembles the shape of a horse's head.

horst   High peaks of igneous rock.

hot pass   Successive welds made over the stringer bead to fill up the groove between joints and to build strength.

Houdry process   Method of catalytic cracking developed by Eugene J. Houdry that revolutionized the refining industry.

hydraulic fracturing   Forcing a formation open by pumping in liquid under pressure.

hydraulic pumping   Using crude oil from the reservoir pumped back into the well under pressure to force more crude to the surface.

hydrocarbons   Petroleum—a mixture of compounds of which the principal chemicals are hydrogen and carbon; crude oil.

hydrocracking   Method of cracking using hydrogen and a catalyst.

hydrogen   A natural element; one of the essential components of which petroleum is composed.

hydrophones   Waterproof microphones used to detect seismic echoes at sea.

hydrostatic head   A difference in height as measured between two points in a body of liquid.

I

igneous rock   Rock formed as molten magma cools.

independents   A person or company engaged in a phase of the petroleum industry that is not a part of one of the larger companies.

Indian Territory   Now incorporated into the eastern part of the State of Oklahoma, it was formerly an area set aside as a home for all the Indians east of the Mississippi.

induction-electrical log   Can indirectly measure porosity and reveal a well's potential as a producer.

inland waterways   The rivers, canals, and intercoastal waterways maintained by the U.S. Government for domestic water transportation.

inorganic theory   A theory of the creation of petroleum that states the elements carbon and hydrogen came together under great pressure under the earth's surface.

integrated company   A large company engaged in many phases of the petroleum industry.

interstitial   Water found in the interstices or pore openings of rock.

isopach   Maps drawn to illustrate the variations in thickness between the correlation markers. This is usually done by shading or coloring.

J

jeep   Electronic device used to detect gaps in the protective coating of a pipeline.

jet sleds   Small, highly maneuverable underwater vehicles powered by jets of water; used by divers.

joint   A single section of pipe.

K

kelly   A hollow 40-foot joint of pipe that has four or more sides with threaded connections on each end to permit it to be attached to the swivel and the drillpipe. Used to transmit torque from turntable to drillstring and can move vertically during drilling.

L

LACT (lease automatic custody transfer)   A tank battery that is fully automated as to recording and shipping into a gathering pipeline.

lateral fault   Fault that has horizontal movement.

lease   A legal document giving one party rights to drill for and produce oil on real estate owned by another. The property described in the document.

lense trap   Reservoir brought about by abrupt changes in the amount of connected pore spaces. Porous oil-bearing rock is then confined within pockets of nonporous rock.

lifts   Various methods of bringing oil to the surface, including pumping units. The pump itself is usually located downhole. The mechanism that operates the pump is often surface mounted and is considered a pumping unit or lift.

light crude   Thinner, freely flowing crude oil of light specific gravity.

lighter ends   The more volatile components or fractions of petroleum. Those with a high API gravity.

limestone   Sedimentary rock composed largely of magnesium carbonate and quartz.

line list   Pipelining instructions that include the names of the owners of the property, the length of the line to be built on the property, and any special restrictions or instructions.

linewalker   A pipeline inspector who walks the length of the line looking for leaks or potentially hazardous situations.

lithology   Study, description, and classification of rocks.

Little Inch   Twenty-inch pipeline, companion to the Big Inch built by the government during World War II.

logging   The lowering of various types of measuring instruments into a well and gathering and recording data on porosity, permeability, types of fluids, fluid content, and lithography.

Lufkin Mark II   Unit having the crank mounted on the front instead of the rear of the beam. It uses an upward thrust instead of a downward one.

M

magma   Rock in its molten state.

magnetometer   Device which detects minute fluctuations in the earth's magnetic field and shows the presence of sedimentary rock.

mandrel Device used to bend pipe without deforming it.

mantle That portion of the earth about 1,800 miles thick between the core and the crust.

marine environment Sediments deposited in the ocean.

marine riser Assembly placed between the ocean floor and a barge or ship on the surface to help maintain the drilling position of the vessel and protect the drillstring.

McAfee process Gulf Refining Co. process of 1915 that used anhydrous aluminum chloride as a catalyst.

Mesozoic Geologic era from about 230,000,000 B.C. to 63,000,000 B.C.

metamorphic rock Created from sedimentary rock subjected to great heat and pressure.

microwaves Electromagnetic radiation with a wavelength between that of short-wave radio and infrared radiation.

middle sample Taken from the middle of the tank.

millidarcies (md) Since average permeability is usually less than one darcy, the measurement is expressed in millidarcies (md) or one thousandth of a darcy.

moonpool Opening in the center of a drillship through which drilling operations are carried out.

mouse hole Shallow hole drilled just to one side of a well in progress to store the next joint of pipe to be used.

mud *See* drilling mud.

mud logger Person who analyzes the cuttings brought up with the drilling mud from the hole.

mud program Planning for the supply of and use of drilling fluids in the drilling process.

multiple completion More than one zone completed from the same hole.

N

naphtha One of the petroleum distillates. Used to make cleaning fluid and other products.

natural gas Gaseous form of petroleum occurring underground.

natural gasoline Casinghead gasoline—a natural condensate of natural gas.

Natural Gas Policy Act of 1978 Legislation that set a new price structure and decontrol schedule for natural gas.

neutron log A nuclear log that can reveal porosity and saturation.

normal fault A fault that has vertical movement.

O

OAPEC Organization of Arab Petroleum Exporting Countries.

Occupational Health and Safety Act of 1971 (OSHA) Extremely comprehensive federal law covering working conditions and health and safety of workers in industry and business.

octane Used to indicate the antiknock quality of gasoline. A fuel with a high octane rating has better antiknock qualities than one with a low rating.

oil pipelines Pipeline designed to carry liquids such as crude oil.

oil string The casing in a well that runs from the surface down to the zone of production.

oil treater Device between the wellhead and lease storage tank to separate natural gas and BS&W from the oil.

olefins One of a class of unsaturated hydrocarbons such as ethylene with many chemical potentials.

organic theory The prevailing theory of the origin of petroleum which states it was formed from plant and animal remains under great pressure beneath the earth's surface millions of years ago.

Organization of Petroleum Exporting Countries (OPEC) A group of Middle Eastern, South American and African states with large petroleum reserves that have joined together to control production and marketing (pricing).

override An additional payment made in excess of the usual royalty.

P

Paleozoic Geologic era from about 600,000,000 B.C. to 230,000,000 B.C.

paraffin The paraffin series is a group of saturated aliphatic hydrocarbons. Paraffin is also used to describe a solid, waxy, hydrocarbon.

pay sands The zone of production—where oil is found in commercially feasible amounts.

perforating Literally punching holes in the casing so the oil and gas can flow into the well from the formation.

permafrost Arctic soil that remains frozen all year.

permeability The factor of a reservoir that determines how hard or how easy it is for oil to flow through the formation. The open space that joins pores of the rock.

pig Device sent through a pipeline to clean it out. *See* go-devil.

pipe gang Part of pipeline crew who line up, prepare pipe, and make initial welds.

pipeline gauger Pipeline company employee who measures the amount and quality of oil entering the gathering lines from the lease tanks.

pipe shoe Device for bending, without deforming, small-diameter pipeline pipe.

pipeyard Working area for cleaning and coating pipeline pipe. Also where two or more joints are welded together before hauling to the pipeline site.

platform The deck or working surface of a rig. An offshore rig anchored to the bottom.

platforming Using a catalytic reforming unit to convert low-quality fractions to those of higher octane.

plugged back To plug off a well drilled to a lower level in order to produce from a formation nearer the surface.

plug trap A salt dome or plug of serpentine rock that has been forced upward by the petroleum accumulated under it.

polymer Synthetic compound having many repeated linked units.

pores The void spaces between the rocks in a reservoir.

porosity The capacity of rock to hold liquids in the pores.

pressure gradient The difference in pressure at two given points.

product Term used to cover any of the products of the petroleum industry.

product pipeline Lines that carry finished products from the refinery.

primary recovery The first period of obtaining oil from a field, by using pumping or free flow.

primary term The period, expressed in time, a lease is written to cover.

prodeltaic plain Geologic environment lying between the normal marine and lower deltaic plain.

public utility Company that sells services to the public such as water, gas, electricity, etc.

pump Mechanical device for lifting oil to the surface.

pumper Person in charge of production and records for a producing well or well field.

pumping Lifting liquids as from a reservoir to the surface by artificial means.

pumping off Recovering the oil that has flowed to the bottom of the well through the formation and then stopping for a rest period until more oil has accumulated. If pumping were maintained continuously, the flow could be stretched so thin the oil would become isolated in pockets and much otherwise recoverable oil wasted.

pumping stations Units placed at intervals along a pipeline to maintain pressure and flow.

R

radioactivity The spontaneous emission of radiation.

radioisotope A particle that has natural or induced radioactivity and which can be introduced into a substance or system as a measuring signal.

rat hole Shallow hole drilled next to a well in progress where the kelly is stored during a trip.

rate of penetration The speed with which the drillstring moves downward.

recovery Obtaining petroleum from a reservoir and bringing it to the surface.

reforming Rearranging the carbon and hydrogen molecules by use of catalysts and heat.

remote sensing Using infrared photography and color television, often from an aircraft, to detect mineral deposits, salt-water intrusion, or faults.

reservoir A rock formation or trap holding an accumulation of petroleum.

reservoir fluid Crude oil, natural gas, and salt water.

residuals Matter left over in boilers or refinery vessels.

residuum The sticky, black mass left in the bottom of a refining vessel.

reverse fault A fault that also moves vertically in the opposite direction of a normal fault.

rippers Devices mounted on crawler tractors to remove rocks from the pipeline trench.

rock cycle The repetitious process of magma cooling into igneous rock and eroded into particles, which become sedimentary rock and then again may be melted into magma.

rod The sucker rod of a pump. A unit of measurement consisting of 16½ feet.

rod pumping Using solid metal rods to lift the oil to the surface.

rotary drilling Using a turning motion to bore into the earth's surface.

rotational fault Of particular interest to the petroleum geologist, a rotational fault moves with a twisting movement.

roughneck Drilling crew member who assists the driller.

royalty Fee paid to the owner of a lease based on the units of production.

run Transferring or delivering from the lease tank battery to the pipeline or tank truck.

running sample Taken by lowering an unstoppered beaker from the top of the oil to the level at the bottom of the outlet and returning it at a uniform rate of speed so that it is about three quarters full when returned.

run ticket Written record of the amount and quality of the run.

Rural Electrification Administration (REA) Government agency formed during the 1930s to bring electricity to rural areas not served by commercial power companies.

S

salt domes Salt plug forced upward through strata because of differences in density.

samples Small amounts of oil drawn from a tank to determine the API gravity and amount of BS&W present.

San Andreas Fault Famous fault line in California running parallel to the Pacific coastline.

sandstone Sedimentary rock composed of grains of sand cemented together by other materials.

saturation The actual amount of fluid available in a given space.

scraper traps Mechanism on a pipeline for inserting and retrieving the pig or go-devil.

scratchers Devices used to clean drilling mud from the walls of the bore so the cement will adhere better.

screen liner Perforated or wire mesh screen placed at the bottom of a well to keep larger particles from entering the bore.

secondary recovery The next attempt at production after all that is economically feasible has been recovered by pumping.

sediment Particulate matter carried along with water which settles to the bottom.

sedimentary deposition The laying down of a layer of sediments in a particular place.

sedimentary rock Rock created under extreme pressure from particles of sediment.

sedimentation The building up of layers of sediment on the bottom of a body of water.

separator Device placed between the wellhead and lease tank battery to separate crude oil from natural gas and water.

seismograph Extremely sensitive recording device capable of detecting earth tremors as used in oil exploration to record man-made shock waves.

semisubmersible Marine drilling rig that can either be anchored to the bottom or maintained at a given position between the bottom and the surface.

shake out   Using a centrifuge to separate any oil that may be present in a sample of BS&W from a well test.

shale   Rock composed of clay and fine-grain sediments.

shale shaker   Mechanical device used to separate bits of shale and rock from the drilling fluid as it comes out of the well.

shell stills   In use by 1870, these horizontal units permit continuous thermal cracking operation.

sidebooms   Crawler tractors with booms mounted on the sides used to lower pipe into trenches.

sidetrack well   A well drilled out from the side of a previous well. Sometimes used to bypass a blockage.

sidewall cock   Valve placed on the side of a tank for the purpose of obtaining small samples.

sidewall sampler   Device used to obtain a sample from the side of a tank.

sidewall tap   Same as a sidewall cock.

single-point mooring system (SPM)   Offshore anchoring and loading or unloading point connected to shore by an undersea pipeline. Used in areas where existing harbors are not deep enough for laden tankers.

slurry   Thin, runny mixture of water and other substances such as clay.

solution gas   Gas dissolved in solution with the oil in a reservoir.

solution-gas drive   If the gas-oil solution is so great no bubbles can form, once the pressure is relieved bubbles do form. As they expand, their pressure drives the oil up.

sour gas   Natural gas containing chemical impurities, notably hydrogen sulfide (H_2S).

specific gravity   The ratio between the weight of a unit volume of a substance compared with the weight of an equal volume of some other substance taken as a standard, usually water.

spot sample   Sample taken at a particular level of the tank.

spread   All of the manpower and equipment necessary to construct a pipeline.

spread superintendent   Person in charge of men and equipment for a pipeline construction job.

spudding in   To begin a new well.

squeeze   To seal off with cement a section of a well where there is a leak either allowing water in or oil out.

squeeze cementing   Process used to fill in any large unwanted openings in the sides of the borehole.

stabilizer   A bushing used on the drillstring to help maintain drilling as close to vertical as possible.

standard pumping rig   Conventional pumping unit using a walking beam to raise and lower the sucker rods.

steam injection   The introduction of steam into the field to obtain secondary or tertiary recovery.

stick welding   Arc welding using a single electrode or rod.

straight hole   A hole drilled with as little deviation from the vertical as necessary.

strapping   Measuring the dimensions of a new tank for the first time to determine its exact capacity.

strata   A layer, as of rock.

stringer bead   The first weld of a pipeline joint.

sucker rod   That portion of a beam pump that actually lifts the oil. The rods are connected to the pump inside the well tubing and to the beam on the surface.

surface lift   Any mechanism at the surface such as a pumping unit.

surf line   That point along a shore where the depth decreases enough to cause the waves to break.

swivel   A rotating attachment point on the bottom of the traveling block.

syncline   Rock formation folded downward.

T

tank table   Chart showing the capacity of a given tank at a given level.

tank battery   Group of storage tanks located on a lease.

tap sample   Sample withdrawn through a sidewall tap or cock.

tariffs   Rules and regulations, including rates, of a common carrier or public utility.

tertiary recovery   The third attempt at production after all the oil has been obtained that is possible by primary and secondary means.

Texas towers   Offshore platforms built during World War II as radar outposts.

thermal cracking   Use of heat to separate petroleum into its various components.

thief   Device used to obtain a small sample of crude from a lease tank.

thiefing   The act of using a thief to obtain a sample.

thief hatch   Opening in the top of a tank through which the thief can be lowered and recovered.

thribble   Three joints of pipe fastened together.

thruster   Mechanism mounted below the waterline of a vessel which can be used to furnish lateral motion.

thrust fault   A horizontal moving fault.

thumper truck   Large truck that by repeated dropping of a heavy weight, can produce shock waves formerly obtained by using high explosives.

tie-in   Connections between pipelines.

tie-in crew   Crew that follows pipeline construction crew to connect the new line to other lines.

tier   Description of type of crude oil (produced in the continental 48 states) for purposes of the Windfall Profits Tax.

toluene   Petroleum derivative of many uses including solvents and explosives.

tool pusher   Person in overall charge at a drillsite.

torsion balance   Device used to measure the gravitational pull of rocks beneath the earth's surface.

torque   Turning or twisting force, as produced by a rotating shaft.

tour   (pronounced "tower") Shift of duty at a well site.

towboat Powerful vessel used to push a string of barges on an inland waterway.

tows A string of barges.

transitional environment Sediments deposited in a delta at the mouth of a river or between two such deltas.

trap A geologic structure that halts movement of a petroleum accumulation.

traveling block The largest pulley on the drill rig. It has the hook attached to its bottom and moves up and down on lines running up to the crown block.

trip Process of pulling string of tools out of a hole (tripping out) or reentering hole (tripping in).

trunk lines A main line. Fed by gathering lines.

Tube and Tank process Standard of New Jersey refining process of about 1920.

tundra Arctic region lying between the permanent ice cap and the more southern forests. The subsoil usually remains frozen and the area supports only limited plant life.

turbodrills A rotary drilling method in which a fluid turbine is usually placed in the drillstring just above the bit. The mud pressure turns the turbine. The drillstring does not rotate; thus, there is no kelly.

two-way spot sample Taken from tanks in excess of 1,000-barrel capacity between the 10 and 15 foot levels.

U

ULCC Ultra large crude carrier—the largest vessels afloat.

unconformity A cap of rock laid down across the cut-off surfaces of lower beds.

undersea pipelines Lines laid underwater on the ocean floor.

upfolds Common deformation of strata—the outer edges are compressed inward and the center rises.

upper sample Taken from the midpoint of the upper third of the tank contents.

upthrust fault A fault that moves vertically upward that may signal the presence of petroleum accumulations.

V

vibroseis Mechanical means of producing shock waves for seismographic exploration without the use of explosives.

viscosity The ability of a fluid to flow.

VLCC Very large crude carrier—a tanker bigger than a conventional vessel but smaller than an ultra large crude carrier.

vug Larger opening between the rocks in a reservoir

W

wagon drills Battery of pneumatic drills mounted on a cart or wagon used as part of the pipeline spread to break up rock.

water and sediment sample Sample taken and "shaken out" in a centrifuge so that it separates into its various components, the amounts of which can be read off directly from a scale.

water drive As water moves in to occupy the space left as petroleum is removed, its pressure forces the remaining oil toward the surface.

waterflooding The injection of water under pressure into a reservoir to drive out more oil. Normally used in secondary recovery.

welding The joining together of metal using a filler metal at high temperature generated by electricity or a gas flame.

welding gang *See* firing line.

well jacket Protective structure, topped with navigational warning devices placed around a completed offshore well.

weight indicator Instrument which constantly displays the total weight of the string in the hole as a well is being drilled.

whale oil Fine, high-quality oil rendered from the blubber of whales.

wildcatter An operator who drills the first well in unknown or unproven territory.

Windfall Profits Tax A 1980-enacted excise tax, taxing producers and royalty owners from 30–50% of the difference in sales price and a government-determined base price, depending on the size of the producer and the oil's tier.

wire line A rope or cable made of steel wire.

wire welding Use of a continuous wire-welding medium supplied on a spool instead of short rods.

workover Cleaning, repairing, servicing, reopening, or perhaps drilling deeper, or plugging back, a well to secure continued or additional production.

Z

zeolite Mineral sometimes used as a catalyst in cracking operations.

zones of lost circulation Crevices, caverns, or very porous formations in which the drilling mud is lost and does not return.

Bibliography

Aalund, Leo R.: "Competition Sparks Refinery Progress," *Oil and Gas Journal*, Vol. 75, No. 35, August 1977.

Anderson, Kenneth E.: *J. B.: His Story—The Biography of J. B. Saunders, Jr.* doctoral thesis, (Stillwater: Oklahoma State University, 1975, n.p.).

Berger, Bill D. ed.: *Facts About Oil*, (Stillwater: Technology Extension, Oklahoma State University, 1975).

Berger, Bill, and Ken Anderson: *Basic Processing Knowledge*, (Tulsa: Penn-Well, 1979).

————: *Gas Handling and Field Processing*, (Tulsa: PennWell, 1980).

Brantley, J. E.: "Percussion Drilling Systems," *History of Petroleum Engineering*, (New York: American Petroleum Institute, 1961).

Burton, Mary Jane: *Science in the Petroleum Industry*, (New York: American Petroleum Institute, n.d.).

Cannon, R. E., and C. B. Sutton: "Four Eras Highlight Gas-Processing History," *Oil and Gas Journal*, Vol. 75, No. 35, August 1977.

Clark, Norman J.: *Elements of Petroleum Reservoirs*, (Dallas: American Society of Mining, Metallurgical and Petroleum Engineers, Inc., 1969).

Dresser-Magcobar: *Drilling Fluid Engineering Manual*, 1977.

Dushman, Sidney: *Chemistry and Petroleum* (New York: American Petroleum Institute, n.d.).

Eckerfield, R. J., and B. Soemantri: "Development Drilling from a Floating Rig," *Journal of Petroleum Technology*, Vol. 29, June 1977.

"Energy is Our Key to Tomorrow," (Washington: American Petroleum Institute).

"Energy Saving Pumping Unit Tested by Cities Service," *Cities Service Today*, No. 7, 1977.

"Exploring," *Conoco 77*, Vol. 8, No. 1, 1977.

Facts About Oil, (Washington: American Petroleum Institute, n.d.).

Feehery, John: "Spindletop . . . Birthplace of a New Era," *Amoco Torch*, Vol. 4, No. 6, Nov./Dec. 1976.

Flowers, Billy S.: "Direct Detection of Hydrocarbons," *Ecolibrium*, Vol. 6, No. 2, Spring 1977.

Gore, Rick: "Striking It Rich in the North Sea," *National Geographic*, Vol. 151, No. 4, April 1977.

Harris, L. M.: *An Introduction To Deepwater Floating Drilling Operations*, (Tulsa: PennWell Publishing Co., 1972).

"Hunting The Atlantic," *Mobil World*, Vol. 41, No. 9, October 1975.

Inghram, E. C.: "The World's Deepest Cable Tool Well," *Drilling*, Vol. 16, No. 10, August 1955.

International Petroleum Encyclopedia 1976, (Tulsa: PennWell Publishing Co., 1976).

Koons, C. B., C. D. McAuliffe, and F. T. Weiss: "Environmental Aspects of Produced Waters from Oil and Gas Extraction Operations in Offshore and Coastal Waters," *Journal of Petroleum Technology*, Vol. 29, June 1977.

247

Leffler, Wm.: *Petroleum Refining for the Nontechnical Person,* (Tulsa: Penn-Well Books, 1979).

Lessons in Well Servicing and Workover: Petroleum Geology and Reservoirs, Lesson 2, (Austin: Petroleum Extension Service, University of Texas, and International Association of Drilling Contractors, 1976).

Lessons in Well Servicing and Workover: Artificial Lift Methods, Lesson 5, (Austin: Petroleum Extension Service, University of Texas, and International Association of Drilling Contractors, 1971).

Loftin, T. D.: "Feasibility of a Fixed Platform for Use in 1300 Feet of Water," *Journal of Petroleum Technology,* Vol. 29, June 1977.

McCain, Wm.: *Properties of Petroleum Fluids,* (Tulsa: PennWell, 1973).

McCray, Arthur W., and Frank W. Cole: *Oil Well Drilling Technology,* (Norman: University of Oklahoma Press, 1959)

"Modern Oil Lifting," *Oil and Gas Journal—Petroleum Panorama,* Vol. 57, No. 5, January 28, 1959.

"The National Energy Outlook," (Houston: Shell Oil Company, 1977).

Oil and Gas Journal, Vol. 75, No. 35, August 1977.

Oil Pipeline Construction and Maintenance, 2nd ed. (Austin: Petroleum Extension Service, University of Texas, 1973).

Price, Paul H.: "Evolution of Geologic Thought in Prospecting for Oil and Natural Gas," *Bulletin of the American Association of Petroleum Geologists,* Vol. 31, No. 4, pt. 1, 1947.

A Primer of Oil Well Drilling, 3rd ed., (Austin: Petroleum Extension Service, University of Texas, and International Association of Drilling Contractors, 1977).

A Primer of Oilwell Service and Workover, 2nd ed., (Austin: Petroleum Extension Service, University of Texas at Austin, 1968).

A Primer of Pipeline Construction, 2nd ed., (Austin and Dallas: Petroleum Extension Service, University of Texas, and Pipeline Contractors Association, 1976).

Progress Review No. 9, Contracts for Cooperative Research on Enhancement of Recovery of Oil and Gas (Bartlesville: Energy Research and Development Administration, 1977).

Ray, Dr. James P.: "Discussing Driller's Second Best Brew," *Ecolibrium,* Vol. 4, No. 4, Winter 1975.

"Rock Properties," *Log Interpretation Fundamentals,* (Houston: Dresser Atlas Division, Dresser Industries, 1975).

Rotary Drilling: Subsea Blowout Preventers and Marine Riser Systems, Unit III, Lesson 4 (Austin: Petroleum Extension Service, University of Texas, 1976).

Seaborg, Glen T.: "Can We Solve Our Energy Problem?" address presented at Golden Gate Welding and Materials Conference, San Francisco, January 1977.

Swigert, Theodore E.: "Handling Oil and Gas in the Field," *History of Petroleum Engineering,* (New York: American Petroleum Institute, 1961).

"U.S. Energy: Yesterday, Today, and Tomorrow," (Washington: American Petroleum Institute).

"VLCCs: The Uncommon Carriers," *Mobil World,* Vol. 40, No. 8, October/ November 1974.

Weaver, Kenneth F.: "The Power of Letting Off Steam," *National Geographic,* Vol. 152, No. 4, October 1977.

Whittaker, Norman R.: *Process Instrumentation Primer,* (Tulsa: PennWell, 1980).

van Poollen, H. K.: *Fundamentals of EOR* (Tulsa: PennWell, 1980).

Index

A

B

C